INSULÆ PHILIPPINÆ
Via Don Padillæ ad Insulas Palaos
14. Novem: 1710.
Samal I.
C. S. Guwam
Mindanao I.
Carragan
1. Ianu. 1711.
Via Reditûs
25. Decemb.
28. Novemb. 1710.
Panlog
Panlog inter Palaos Insulas Maxima
Sarol
Falivelap
Comul
10. Decemb. 1710.
I. Sonsorol als S. Andreæ
Seu Losacay
Farale seu Poulo I.
Merrieres I. seu Ontocoso
Piguela
Linea Æquinoctialis.
Los Papuas
Terra dos Papous
Habitans des Iles Pelew

In Memory of
President Ngiratkel Etpison

and to
Yuang Ngirabiol Etpison

In Memory of
President Ngiratkel Etpison

and to
Yuang Ngirabiol Etpison

Insulæ
Guaham I.
Marianæ
Fayo
Old
Pisarap
Lamuet
Pis
Mago
Faravolvol
Falato
Apigap
Olutil
Falippe
Tavas
Olor
Ett
Pie
Farvile
Laf
Feley
Boo
Carvolvol
Yap
Paiz
Amorsot NB.
Tolobas
Coop
Olotup
Pug
Loto
C
Fataray
Pitamgarus
Falalap
B
Mayaya
Magomer
Yassor
Luxep
Pariali
Alangari
Ulimaray
Tavr
Culaot
Sabo
Pogol
Eu
Corobol
Fagatale
Sabolocolop
Pirigarava
LÆPA= NOVÆ PHILIPPI= NÆ LAOS
87.
Aliàs
AGUS S. LAZARI vel CAROLINÆ
Pauli Klain è Soc: IESU è in Philippinis Insulis et Theol: Professoris
ilæ, ac ex itinerario illustris viri Don Francisci Padilla
Faratay
Salva
Otoc
May
Torres
Faramalo
Farali
Faramit
Farilies
Falu seu
Lamurec
Sedes Regis
Pata
Alingaray
Tacaylap
Raor
Marioc
Ulier
Parivu

Photo credits:

Kevin Davidson:
Page 3, 21 top & bottom right, 22, 35 top, 45 left 2, 46, 47, 48, 53 top 2, 66 top, 86, 99 bottomleft, 131, 132, 227 top, 228 bottom 232 bottom 2, 235, 236 insert, 237, 238, 251, coverbox front.

Patrick L.Colin/ Coral Reef Research Foundation:
Page 12 top, 18 top, 20 bottom, 35 bottom, 37 airport, 41, 43, 52, 54, 93, 108 bottom, 173 top, coverbox back.

Mandy Etpison:
Page 5,6, 7, 9, 25, 26 middle, 27, 28, 29, 31, 32 top, 34 bottom, 36, 37, 38, 39, 40, 44, 51, 53 bottom 2, 58, 59, 60, 62, 63, 64, 65, 66 bottom, 67- 92, 94, 96, 99, 134, 135 bottom right, 139, 140, 141, 142, 148, 149, 163, 171, 172, 173, 174, 175, 176, 178- 189, 190 top 2, 191, 193, 194, 195, 196, 199, 200, 201, 202, 203, 204, 205, 206, 207, 208, 209, 210, 211, 212, 214, 215, 216, 217, 218, 220, 221, 222, 223, 224, 226, 228 top, front cover.

Tim Rock:
Page 6 left, 14, 15 bottom2, 16, 17 bottom2, 18 bottom 2, 19 top right, 20 top, 21 bottom 2, 23, 24, 26 top & bottom,, 29 bottom left, 30, 32 bottom.

William E.Perryclear:
Page 4, 34 top, 55 bottom 4, 56, 57, 79 3rd left, 135 left 2, 190 bottom 2, 192, 239 bottom, 243 top, 247 bottom left, back cover.

Bert Yates: Page 45 right 2, 49, 50,110, 136, 223 top left, 230 top. **Tanya Brungard:** Page 177. **James Watt:** Page 12 bottom. **Jolie Liston:** Page 77 top. **Jo Anne Van Tilburg/ David C.Ochsner:** 85 top left, 87 top. **Palau Bureau of Arts and Culture:** 89 left 2. **Thomas Lautz:** Page 99 black and white. **David B,Fleetham/ Seapics.com:** Page 197 top, 232 3rd. **Henni Rall:** Page 221 **Dr. Victor Yano:** 25 top, 29 top left. **National Archives- Naval Operational Archives:** 129, 131 top. **Etpison Museum collection:** all old prints and maps. **Herbert W. Krieger:** 10, 15 top. **Walter v. Rummel:** 137 top. **Hamburg Sudsee-expedition 1908-1910:** *(Augustin Kramer, Palau 1-5)*: Page 97 top, 121, 123, 126, 127, 128, 137, 138, 143, 144, 145, 146, 147, 150, 164, 165, 166, 167, 170, 197 bottom. **(Muller-Wismar, Yap):** 100. **Mike Veitch:** Page 236 background. **Doug Schloss:** Page 252. **Del Benson:** Page 254.

Staatliche Ethnographische Sammlungen, Sachsen- Museum fur Volkerkunde, Dresden, Germany:
Page 155 bottom, 156 bottom, 158 bottom- Eva Winkler, 162.

Staatliche Museen zu Berlin- Preuβischer Kulturbesitz, Museum fur Volkerkunde, Germany:
Page 151, 155 top, 157 top- Martin Franken; 168- MV Fotoatelier/ Dietrich Graf, 169- W.Schneider-Schutz/ Dietrich Graf.

British Museum:
Page 105, 113, 152 top, 156 top, 157 bottom 3, 158 top,159, 160, 194 2nd, 195 middle, 198.

Acknowledgements:
Special thanks to: The Belau National Museum for allowing me to photograph their collection; Udo Horstmann for arranging all the wonderful images from the German collections; Alison Deeprose; Markus Schindlbeck; Petra Martin and Irene Godenschweg. As always to my boys Tkel and Iked, and to Shallum, I love you guys, and thanks for putting up with my long hours on the computer these last months...

Published by/ Please send any inquiries to:
Tkel Corp. P.O.Box 7049, Koror Palau PW 96940 Email: etpison@palaunet.com
ISBN 982-9086-02-X

Contents

Micronesia

Okinawa
Iwo Jima
China
Taiwan
Hong Kong
Parece Vela
Northern
Marianas
Philippines
Vietnam
Saipan
Tinian
Rota
Guam
Samar
Micronesia
Yap
Ulithi
Fais
Palawan
Cebu
Mindanao
Palau
Ngulu
Sorol
Lamotre
Woleai
Eauripik
Satav
Borneo
Kepulauan
Talaud
Southwest Isl.
Western Caroline Islands
Morotai
Halmahera
Sulawesi
Equator
Malaysia
Pulau Waigeo
Kepulauan Schouten
Manus
Admiralty
Melanesia
New Guinea
Bismarck
Archipela
Indonesia
Java
Bali
Pulau Alor
Pulau Wetar
Timor

BELAU ER

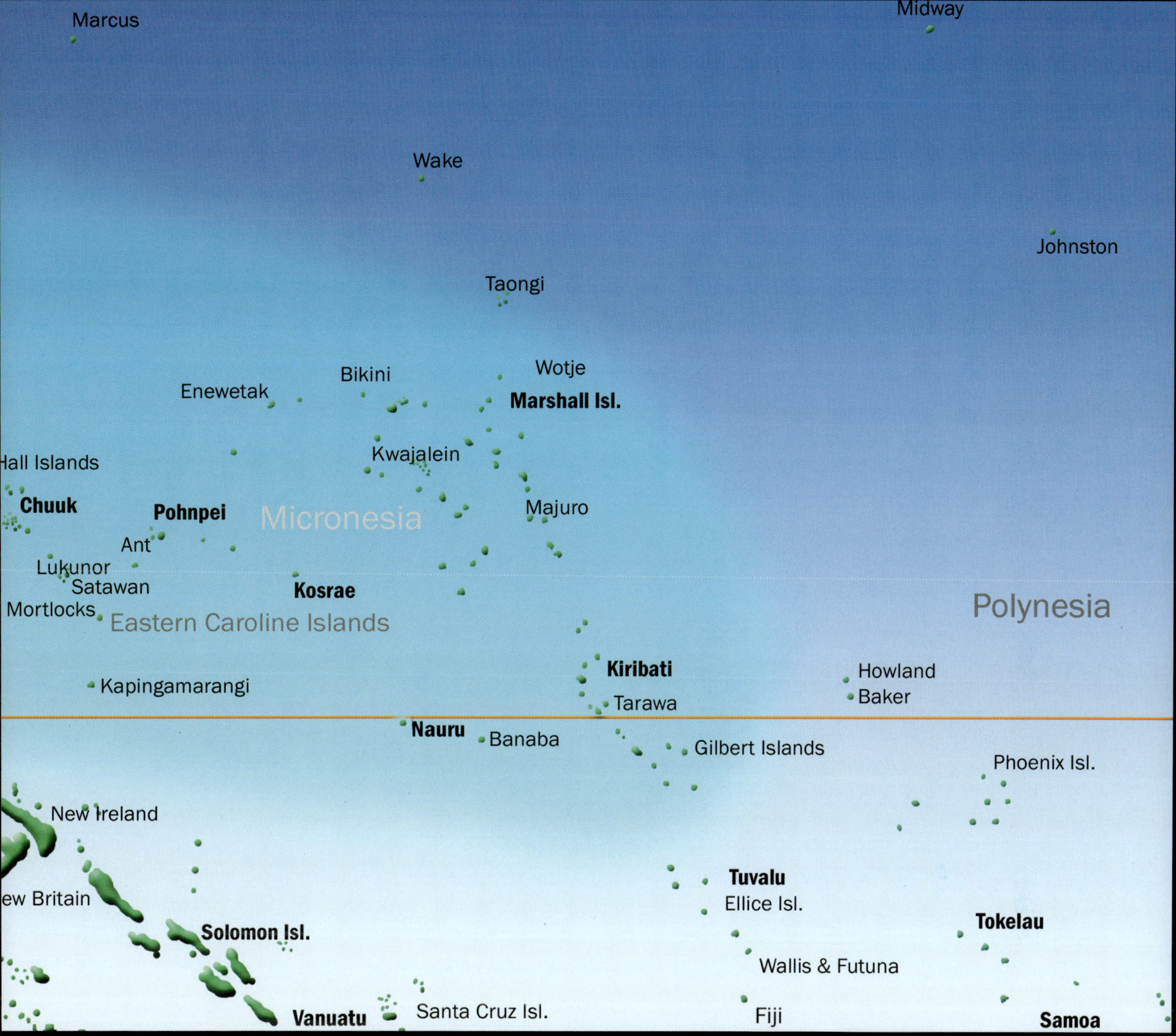
Marcus
Midway
Wake
Johnston
Taongi
Wotje
Bikini
Enewetak
Marshall Isl.
Kwajalein
Hall Islands
Chuuk
Pohnpei
Micronesia
Majuro
Ant
Lukunor
Satawan
Kosrae
Mortlocks
Eastern Caroline Islands
Polynesia
Kiribati
Howland
Baker
Kapingamarangi
Tarawa
Nauru
Banaba
Gilbert Islands
Phoenix Isl.
New Ireland
Tuvalu
Ellice Isl.
ew Britain
Solomon Isl.
Tokelau
Wallis & Futuna
Vanuatu
Santa Cruz Isl.
Fiji
Samoa

Micronesia

Micronesia, which means little islands in Greek, is a widely scattered group of coral atolls and islands located between the Philippines and Hawaii in the western Pacific, each with their own unique culture and history. It includes seven political entities: the Republic of Palau; the Federated States of Micronesia, made up of the island chains of Yap, Chuuk, Pohnpei and Kosrae; the Commonwealth of the Northern Mariana Islands; the non-incorporated Territory of Guam; and the Republics of the Marshall Islands, Nauru, and Kiribati. Kiribati includes the far-flung Phoenix, Line and Gilbert Islands. Some of these have little in common with the rest of Micronesia, and are culturally more Polynesian.

The annual temperature range is very narrow, and temperatures rarely fall below 26° C. Humidity is very high in the west (80%) and moderate in the east (40%). During winter months strong typhoons occur, and hit certain islands regularly, causing a lot of damage on the lower coral islands. Flora and fauna is limited in variety, but unique, and most of the Micronesian islands have endemic species of birds and reptiles.

The more remote outer islands are still not much affected by tourism and development, and aside from some imported materials and utensils life goes on as it has for centuries. The main islands with international flight connections are now rapidly changing though. Some of the islands also did not escape the violent effects of foreign colonization and war, and have little left of their original culture, such as the Northern Marianas, Guam, and some of the Marshall Islands.

After European traders and whalers brought the first firearms and diseases in the 1800s to most of Micronesia, the Spanish missionaries tried to convert them by force and destroy their local cultures; the Germans followed, leaving the islanders mostly alone, and only caused major trouble on Pohnpei. After that Japan took over, and in the 1930s some of the islands looked like Japanese towns, with more Japanese living there than locals. The islanders then got caught between the Japanese and the Americans during WWII, a conflict which had nothing to do with them. Some of their main islands were bombed flat by the Americans, and thousands of natives were killed and displaced from their villages. After WWII the Marshallese were used as Guinea-pigs for atomic bomb experiments by the Americans, and on other islands big stretches of land were confiscated by the Americans for military use and bases, some of which are still there today. Looking at this Micronesian history in a nutshell, it is a miracle some of these islands have been able to hang onto their unique culture and still have strong traditions.

Most of the Micronesian islands originally had complex matrilineal societies, ruled by titled Chiefs. Some islands had independent village communities with many Chiefs and equal standing, while others had important districts that received tributes from other districts with lower social standing, and were ruled by a few aristocratic Chiefs. Today most of the main Micronesian islands have gone from the Stone Age to Computer Age in a time span of a hundred years, with young kids from outer islands going off to US colleges, many not returning, while their grandfathers are still walking around in loincloths, and are trying to hold on to their old traditions and way of life.

Early 1900s German drawing, showing different people of the Pacific Islands.
Australians: 1. North Australia, 2. South Australia, 3. West Australia, 4. Tasmania
Melanesians: 5,12,13. Bismarck Archipelago, 6. New Caledonia, 7. Solomons, 8. Vanuatu, 9. New Guinea, 10/11. Fiji.
Micronesians: 14. Yap, 15. Mortlock Isl., 16. Gilbert Isl., 17/18. Palau, 19. Caroline Isl., 20. Chuuk, 21. Marshalls
Polynesians: 22. Tonga, 23. Hervey Isl., 24. New Zealand, 25. Marquesas, 26/27. Nauru, 28. Tuamotu, 29/30. Samoa

After WWII most of the islands were kept under wraps, and tourists discouraged from visiting, under a US Trust Territory. This allowed the United States to use the islands for military purposes and authorized it to deny entry to outsiders. In return the US was to further modern education and sanitation, and promote the islands toward self-government and independence.

Disputes that previously would have been fought out in wars between villages and islands are today fought out in the political arena and the courts, and the local politics of most Micronesian islands are complex and frustrating to deal with, being a odd mixture of western and traditional bureaucracy, run by elected government officials as well as traditional Chiefs. The current political status of some of the islands is still fragile. The Federated States of Micronesia set-up for instance is far from ideal, combining four far-flung island chains that are culturally very different, and several of these islands are already looking at ways to change their situation, and want to become independent.

Although they are not very well-known, or maybe because of that, the Micronesian islands offer great holiday destinations for nature-lovers and scuba divers, and shoppers can find some unique handicrafts. Be aware that visiting some of the remote outer islands needs special permission ahead of time, and transportation is not always reliable. Accommodations there are not exactly 5 star, and you are expected to bring some gifts. Always read up on the culture before you visit, and respect and follow local customs, the people there will appreciate it.

I once flew all the way to Woleai, only to find out that women are not allowed in the village or able to talk to the Chiefs. Luckily my husband was with me, who, while I sat at the airstrip, gave out presents (cigarettes, betel nuts and bread) and was invited to drink coconut wine with the local Chiefs. He asked permission for me to come into the village and take pictures, which they later, quite cheerfully after about a gallon of coconut wine, allowed.

Micronesian people in general are a proud people, and consider themselves equal if not above any foreigners, which is sort of refreshing compared to the colonial attitude you encounter in the Philippine and Indonesian islands, where they were taught to look up to anyone foreign. A few of the island cultures were totally ruined by missionaries and diseases, but most Micronesian cultures have survived because they pretended to accept foreign rule and religion, but in reality only did so as long as these didn't interfere with their local beliefs and traditions. Although most of the main islands may seem very westernized, the people are islanders who still deep-down believe in magic and spirits, and follow a lot of the traditions of their ancestors. Don't be surprised about a Christian senator with a college degree consulting a traditional medicine man, or to see him painting his house with leaves, because he thinks someone put a spell on him. These things still happen in Micronesia.

Their weakness is their laid-back attitude and inability to look ahead and plan for the future. The main priorities in life seem to be to have a car, TV, boat and air-conditioning, the rest of life they sort of take as it comes. Their time is consumed with local politics and disputes between clans, villages and Chiefs, and they have always been so busy fighting amongst each other, that they have never been able to stand together as one people, even within the individual island chains.

This is a book on the people of Palau, but I want to start out with a brief glimpse of the other islands, to give you an idea of the amazing variety of islands and people that together are Micronesia. Each island at the time of the first European visitors had their own unique traditional attire and culture, even the type of grass skirts between the island chains were not the same. On some islands like the Marshalls the men wore grass skirts and the women woven mats, while in the Gilbert Islands the men wore the woven mats, while the women had grass skirts. Men on Yap and Chuuk wore loincloths, on Guam and Palau they were totally naked. On some islands the men did the fishing and the women planted, while it was the other way around on other islands.

Some of the most unusual money systems in the world are found on Yap with their enormous stone discs and shell money, and on Palau with its antique moneybead system. People from the outer Caroline and Marshall islands were expert navigators and fishermen, skills necessary for them to survive and trade with other islands, which they did long before the first Europeans arrived in 1521. Simple adzes made from shell, bone and wood enabled them to make houses, canoes and wooden utensils with amazing ingenuity and skill. The ancient rock drawings and stone monoliths of Palau, the Latte stones of Guam and the Marianas, and the stone city ruins of Pohnpei and Kosrae make you wonder about the very early settlers of these islands, where they came from, how they arrived there, and how they lived.

In general it is believed that the Micronesians are of Asian origin, and that people arrived from Philippine and Indonesian islands to Palau, Yap and the Marianas around 1,300 BC. A second group of settlers is believed to have come from eastern Melanesia around 500 BC to settle the central and eastern islands. Intentional internal migrations also occurred, like the Yapese people who settled on the Southwest Islands of Palau.

Nauru

Other names: Pleasant Island **Political status:** British Commonwealth since 1999

Population: Reduced to 900 in 1886 after civil war. Currently over 11,000.

Attractions: Single island only 20 square km, good for a one-day visit.

Pre-contact: The people were of Polynesian and Micronesian descent, divided up into 12 matrilineal clans, named after marine life. Each clan had its own songs and crafts.

Historic highlights: High-grade phosphate was found on Nauru in 1900, which Germany, Japan, England, Australia, and New Zealand fought over till Nauru became independent in 1968. Nauru was a UN Trust Territory under Australia after WWII. Its people were among the richest in the world for a while after 1968, but the phosphate resources are expected to be exhausted in the next 2 decades, and the island is an environmental disaster. Nauru filed suit against the Australian government for the mining damages, and it was settled out of court in 1993 for $109 million Australian dollars.

Kiribati

Political status: Independent Republic since 1979. The Polynesian Ellis Islands, formerly part of Kiribati, formed their own independent Republic of Tuvalu in 1975.

Capital: Tarawa **Population:** Over 85,000.

Attractions: Most unspoiled islands in Micronesia, largest atoll state in the world, made up of 33 coral atolls, widely scattered between Banabe and Christmas Island. Three main island groups, 16 Gilbert Islands, also called Tungaru, 8 Phoenix Islands, 8 Line Islands, and the single phosphate island of Banabe, also called Ocean Island.

Pre-contact: Settled around 3,000 years ago. Fierce warriors, known for their shark teeth daggers and swords, and body armor made from woven coconut fiber. Ruled by Chiefs.

Historic highlights: First detailed charts made in 1840. Whalers, traders and missionaries visited since mid-1800s. The Gilbert and Ellis Islands were claimed by Britain in 1892, and the phosphate-rich Banabe in 1901. The Line and Phoenix Islands followed. Some islands were occupied by the Japanese during WWII, and the US invaded Betio Island in 1943. The islands were under the British High Commission after WWII.

Marshall islands

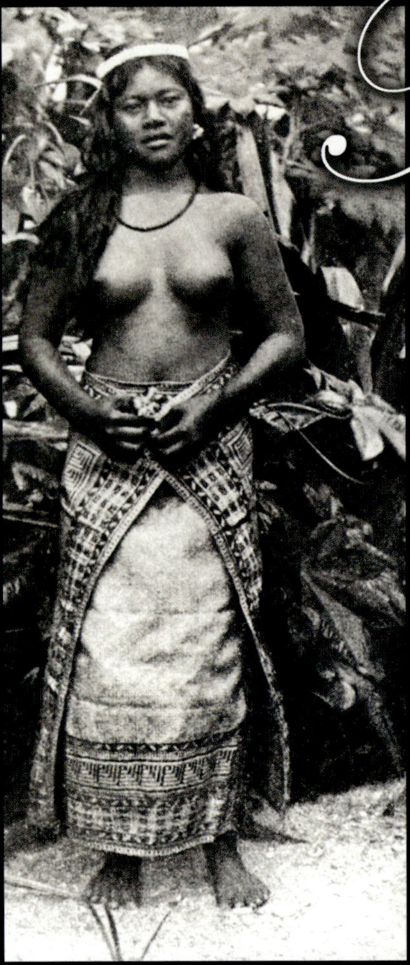

Political status: Self-governing democratic Republic, under a Compact of Free Association with the United States. Independent since 1986.

Capital: Majuro

Population: Pre-contact unknown. Currently around 60,000, the majority living on the Majuro and Kwajalein atolls, which are two of the most densely populated atolls in the world.

Attractions: Over a 1,000 islands, grouped together in 29 coral atolls and 5 coral islands, widely scattered. Scuba diving, (reef sharks and wreck diving), fishing, WWII sightseeing, handwoven crafts and baskets, stick charts, pristine outer islands.

Pre-contact: Earliest settlement evidence 1960 BC. The Marshallese people were fishermen, canoe builders, and expert navigators using stick charts. Marriages were planned for, and wars were fought over land control. Women wore long woven mats, and men thick grass skirts. It was a matrilineal society, with Chief titles descending through the mothers clan.

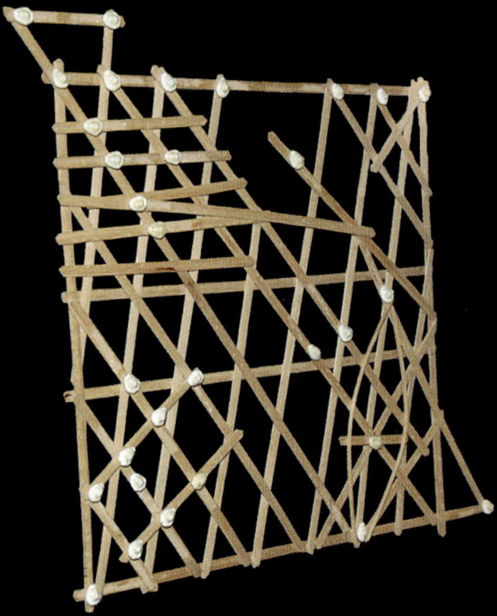

Historic highlights: The islands were named after the British captain John Marshall, who visited several of the islands in 1788. The Russian explorer Otto von Kotzebue described the islands in 1817 and 1825. Whalers were discouraged by the islanders. After a period of colonization by Germany and Japan since the 1880s, it became a U.S. Trust Territory after WWII. The Japanese had built military bases and used the islands to set up an invasion of the Gilbert Islands in 1941. The US invaded Kwajalein and Enewetak in 1944, and bombed all of the other atolls with Japanese bases. From 1946 to 1958 the United States started atomic bomb experiments on Bikini and Enewetak atolls. Some islanders died from radiation-related ailments, while others have lingering health problems even now. A total of 67 nuclear bomb tests were performed during post-war years. The US Congress has allocated $240 million dollars recently to compensate victims, after several lawsuits were filed against the US government.

The capital, Majuro, is very westernized, but some of the outer islands are more pristine with only simple accommodations. The island of Kwajalein is leased to the United States military for missile testing, and is off-limits to most visitors. The local economy almost fully depends on US funding. Bikini has recently opened up to scuba divers, who are attracted by the ship wrecks.

Left, center: Stick navigation charts were made traditionally by the Marshallese out of pandanus root and cowrie shells. The geometric arrangement of the sticks and shells showed wave patterns, stars, and island locations. Navigation was taught to youngsters this way, who had to memorize the charts starting at age 6, and it passed down through generations.

Aerial view of Enewetak atoll

Grey reef sharks around Bikini atoll

1828 drawings of Kosrae.

Kosrae

Previous names: Strong's Island, Ualan, Kusaie.

Political Status: One of the four Island States of the Federated States of Micronesia, together with Pohnpei, Chuuk, and Yap, since 1978.

Capital: Tofol

Late 1800s drawing of a Kosraean man.

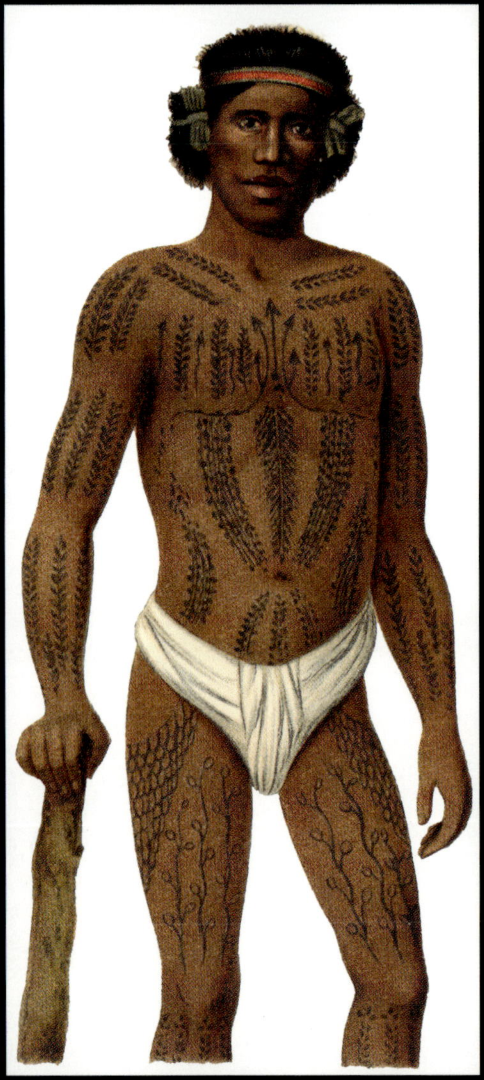

Population: Pre-contact estimated at 5,000, brought down to 300 in the 1800s because of foreign diseases. Currently around 8,000.

Attractions: A single, high volcanic island, the only State without outer islands. Ancient ruins of a stone city, canoeing.

Pre-contact: Ruled by one Paramount Chief. Traditional wear short woven skirts for women, and loincloths for men. Matrilineal society.

Historic highlights: First real European contact with the French ship Coquille in 1824. During the 1800s Kosrae became a well-known whaling port, and the few Kosraeans still alive were converted by Protestant missionaries, and their culture virtually obliterated. The ship Leonora from American trader Bully Hayes sank in Utwe Harbor in 1874, and became a famous shipwreck, rumors about a buried treasure somewhere on Kosrae have circulated ever since. The Japanese developed the island before WWII, and took over several local villages. Until 1977 Kosrae was part of the Pohnpei district under a US Trust Territory. The island demanded separation and later became a separate State under the Federated States of Micronesia. Kosrae is the most religious island in Micronesia, 90% of the people are Congregationalist, and almost everything is closed on Sundays. The traditional matrilineal society has now changed into a western-style patrilineal system. Almost nothing of the traditional culture survives today.

Lelu ruins

The ancient stone city of Insaru on Kosrae's Lelu Island consists of over a 100 compounds, some surrounded by stone walls up to 21 feet high. These ruins, located on the East coast within the reef, were built on artificial fill, and were the administrative and religious center of Kosrae between 1250 and 1800 AD. The King and the High Chiefs lived on the island, and were supplied with food from the main island. The stone city also had tombs where deceased royalty were mourned. Their bones were later dropped in a hole on the reef. The city was built with natural prismatic basalt logs, which were stacked together with volcanic rocks.

Around 1800 a severe typhoon hit the area, damaging the sea walls and eroding the shore. The resulting famine and foreign diseases brought by whalers in the 1800s caused a dramatic population decrease from 6,000 to around 300 people in a few decades. This marked the end of the complex traditional Kosraean hierarchical society, which had a King, High Chiefs, Low Chiefs, and commoners.

Today little of the original Lelu stone city remains, since most of modern Lelu has been built right on top of the old ruins. Local legends say Lelu city was built by a Chief exiled from Pohnpei. His son later returned to Pohnpei with an army of hundreds of men, and conquered Nan Madol to establish new districts and Chiefs there.

1943 photograph

1828 drawing

Pohnpei

Sokehs Rock, a signature landmark.

Previous names: Ascension Island, Senyavin, Ponape. The name Pohnpei means "upon a stone altar".

Political status: One of the four Island States of the Federated States of Micronesia together with Kosrae, Chuuk, and Yap, since 1978.

Capital: Palikir Island is the capital island of the Federated States of Micronesia. Kolonia is the capital of Pohnpei.

Population: Pre-contact estimated at 30,000. Dropped to less than 5,000 by 1860 because of smallpox. Currently around 35,000, 95% of those are living on the main island.

Attractions: The third largest volcanic island in Micronesia, with 8 outlying atolls. Famous for the ancient stone city ruins of Nan Madol, Waterfalls, hiking, scuba diving, kayaking.

Pre-contact: Traditional wear for both men and women were hibiscus fiber skirts, usually dyed yellow. Woven decorated belts, necklaces, and carved dancing paddles showed rank and status.

History highlights: Coastal areas were inhabited as early as 100-200 AD. Not much is known before the Saudeleur Kings took over around 1200. After they were conquered by Kosraeans, Pohnpei was divided into 5 districts, each with two ruling families. The island was claimed for Spain in 1595. Whalers and missionaries arrived in the mid-1800s. A smallpox epidemic in 1854 killed over 2,000 Pohnpeians. After the Spanish, the Germans arrived in 1899. The Sokehs Rebellion of 1910 happened when a Pohnpeian workman killed a German overseer who beat him. The Germans exiled over 400 Pohnpeians to Palau to avoid further incidents, and brought in other Micronesians for the forced road-building on Pohnpei. The Japanese took over in 1914. By the 1930s there were over 10,000 Japanese and Koreans living on Pohnpei. Kolonia was levelled by US aerial bombings in 1944, and became a US Trust Territory after WWII.
The people from the outer atoll of Kapingamarangi are Polynesian, and many moved to Pohnpei after a bad typhoon in the early 1900s. They are known for their wood carving skill.

Left, bottom: Dances are still being performed by whole families at special occasions on Pohnpei.

Sakau

Sakau is made from the roots of the Indian pepper plant *(Macropiper methysticum),* is very strong, and slightly narcotic. The drink is known as Kava elsewhere in the Pacific and Polynesia. Traditionally it was only drunk on Pohnpei during religious ceremonies with the High Chief, with strict serving based on rank. It is held with both hands, and eyes closed. The root is pounded on a flat stone, and the pulp squeezed by hand through hibiscus bark fiber. The juices are mixed with some water, and served in a coconut shell which is passed around. Your mind seems to stay clear after drinking too much Sakau, but your body refuses to move properly. In Micronesia, Sakau is only made on Pohnpei and Kosrae.

Nan madol

The ruins of Nan Madol on Pohnpei are similar to, but even more impressive than those of Lelu Island on Kosrae. The ruins consist of 92 artificial islets, inter connected by waterways, and surrounded by stone walls. Located on the southeast coast of Pohnpei, the 200-some acre city was constructed using prismatic basalt quarried from the main island, and coral. Most activity seems to have taken place there from 800 to 1300 AD. The city housed the royalty, their tombs, and had areas for specific religious and ceremonial purposes.

One islet had a well where they used to keep a sacred eel, which was fed with turtle intestines during special ceremonies. All the weapons were kept on this same islet. Turtles were kept on another islet in an artificial pond.

After 1200, Pohnpei was ruled by a Paramount Chief called Saudeleur. Nan Madol flourished, and more stone walls were built, some up to 25 feet high. Activity went down after 1600. The largest and most impressive wall can be seen on the islet of Nan Douwas, which has an inner courtyard and central tomb, which housed the remains of the last Saudeleur. After the Saudeleurs were overthrown by an exiled Chief and Kosraeans, Pohnpei was divided into 5 districts with a ruling system of ranked Chiefs.

Left: An antique Pohnpeian carved ceremonial dancing paddle. These paddles and woven decorated belts were important possessions and signs of rank for Pohnpei people.

Next page: An aerial view of Nan Madol, and details of the huge walls. Although mostly overgrown and not regularly cleared, the different islets can still be seen clearly.

Below, right: The only petroglyphs on Pohnpei are found on a large boulder.

Below, left: Pohnpei waterfalls

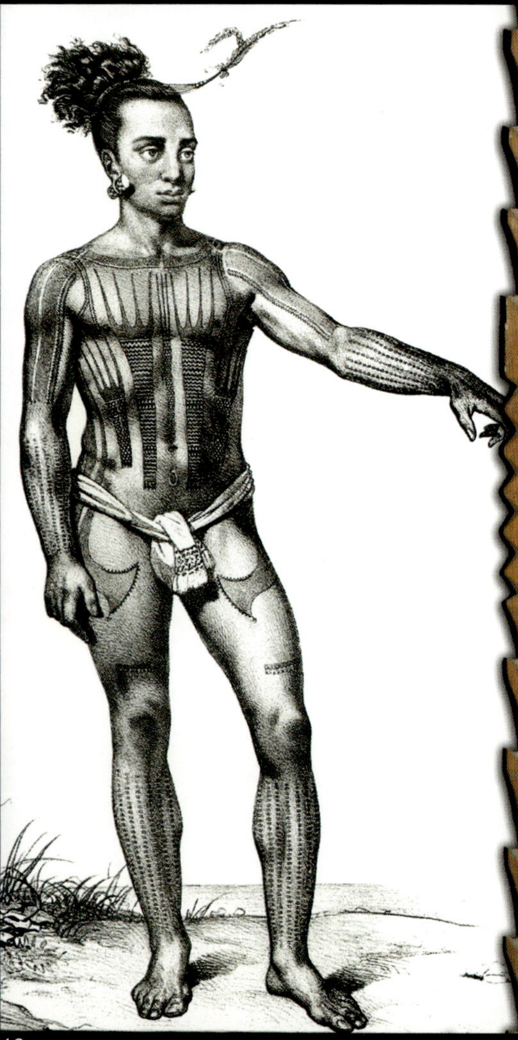

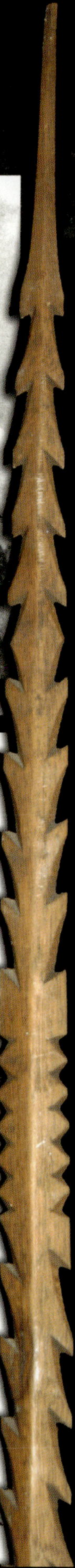

Chuuk

Previous names: Truk. Chuuk means "mountain"

Political status: One of the four States in the Federated States of Micronesia with Yap, Kosrae and Pohnpei.

Capital: Weno, formerly known as Moen.

Population: 57,000, one third living on Weno.

Attractions: Includes 15 main islands, and the 200-some outer islands of the Mortlocks, Hall Islands, and Western Islands. The Truk Lagoon is a famous dive destination because of its sunken Japanese WWII wrecks.

Pre-contact: Settled around 1300, perhaps by Kosraeans. Islands ruled by clans. Traditional wear grass skirts or woven skirts for women, loinclothes and cloaks for men, depending on the area. Shell jewelry.

Historic highlights: The Spanish traded sea cucumber since 1814. The islands were taken by Germany in 1899, followed by the Japanese in 1914. Chuuk became the central Pacific base for the Japanese Imperial fleet during WWII. In 1944 the US bombed the entire fleet for 2 days, sinking around 60 ships.

Left and right: Love sticks were used by courting males. Individual designs identified a boy to a girl when he pushed his stick into the girl's hut at night, and she would choose to see him or not. Concrete houses put an end to this interesting tradition.

The outer Islands were known for their elaborate tattooing and ear and hair decorations, as shown here in early photos and a 1828 drawing. The remote Western Islands share close ties with the Yap islands.

Aerial view of the Lagoon Islands of Chuuk.
Weno town, Chuuk.

Truk lagoon

A tank sits on the deck of the Nippo Maru, one of over 60 Japanese ships sunk by US aerial bombing in 1944. These wrecks bring thousands of scuba divers to Chuuk every year from all over the world.

Japanese field artillery on the deck of the Nippo Maru.

Many Japanese remains are still inside the ship wrecks. It is prohibited to take souvenirs.

Guam

Previous names: Guahan

Political status: Unincorporated US territory.

Capital: Agana, also known as Hagatna

Population: Pre-contact estimated 50- 100,000, by 1710 it was reduced to 3,500 because of warfare and diseases. Currently over 150,000, only 40% of those are Chamorro.

Attractions: Largest single island in Micronesia. Known for the stone Latte column rows. A total of 39 sets is known from Guam. The Latte are also found on the Mariana Islands.

Pre-contact: Settled before 1500 BC. Traditionally the Chamorro men went naked, and the women covered themselves only with leaves. It was a matrilineal clan society.

Historic highlights: Spain claimed Guam and the Northern Marianas in 1565, and it became a busy stop-over for Spanish galleons on the trade route between Manila and Acapulco. In 1668 the first Catholic mission was established, and priests tried to change the culture, causing rebellions and warfare. By the mid-1800s, whalers frequented the area. The US Navy took over the islands in 1898. The Japanese invaded Guam in 1941, and the Americans invaded and took it back in 1944. Large tracks of land were confiscated by the US in preparation of the invasion of Japan, and turned into permanent bases after the war.

Part of Guam still belongs to the US military, with naval and airforce bases. Tumon Bay is a commercial area with duty-free malls and high-rise hotels for the mostly Japanese tourists. The island is hit regularly by typhoons and earthquakes. Supertyphoon Paka in 1997 recorded 235 miles per hour winds, and an earthquake measuring 8.1 on the Richter scale shook the island in 1993. Guam is the main access point for most flights to and from the other Micronesian islands.

Northern marianas

As Nieves Latte stone quarry on Rota.

View over Rota.

Previous names: Ladrones (Islands of Thieves), Las Marianas

Political status: US Commonwealth since 1978.

Population: Pre-contact estmated at 50,000. Reduced to less than 4,000 by 1914. Currently around 60,000, 90% of them living on Saipan.

Attractions: Made up of 14 islands. The Mariana Trench that runs east of the islands is 38,635 feet deep, the greatest ocean depth. Active volcanos are found on the northern islands, the last eruption took place in 1981 on Pagan. Saipan has 5 golf courses, while Tinian has a big casino, and the largest Latte stone formation, called "the House of Taga", (see above 1742 European drawing). The island of Rota has a Latte quarry called "As Nieves", right, with unfinished Latte still in the ground, and a total of 46 Latte sites. Most Latte activity took place between 1350 and 1650 AD.

Pre-contact: Settled around 1500 BC by Chamorros.

Historic highlights: Spotted as early as 1521, the islands were claimed by Spain in 1668. The native people were moved away to Guam in the 1690s to be converted. Around 1820 outer islanders from Yap and Chuuk started settling on the abandoned Mariana Islands. By 1885 some of the original Chamorros moved back from Guam. Germany took over in 1899, followed by the Japanese in 1914, who brought in some 45,000 workers from Okinawa to work on the cane fields by the 1930s. The US invaded Saipan and later Tinian in 1944, and used Tinian to stage the atomic bomb missions to Hiroshima and Nagasaki in 1945.

The CIA closed off part of the Marianas from visitors from 1948 to 1962. Little remains of the traditional Chamorro culture today. In 1989 the Manila galleon Nuestra Senora de la Concepcion was discovered, which sank off Saipan in 1638.

The largest stone money sits on Rumung Island It is still there today, but the island is closed to visitors.

Yap

Previous names: Waab **Capital**: Colonia

Political Status: One of the four States of the Federated States of Micronesia with Chuuk, Kosrae and Pohnpei, since 1978.

Population: Pre-contact estimated at 50,000. By 1914 reduced to less than 6,000 because of leprosy and influenza. Currently over 11,000.

Attractions: Consists of the 4 islands of Yap proper, and 9 outer island atolls. Known for the large traditional stone money discs. Annual Yap day on March 1st shows the 4 traditional dances. Traditional meeting houses, scuba diving with Manta rays. Colorful flower leis and marmars.

Pre-contact: Traditional wear for women were grass skirts on Yap, and woven skirts on the outer islands, loincloths for men. Aside from the stone money, shell necklaces were used for rank and money. Yap received yearly tributes from the outer islands, who were expert canoe builders and navigators. Full intricate body tattooing was common.

Historic highlights: Contact between Yap and the outside world was always limited. Ulithi was first visited in 1525 by a Portuguese ship. The outer islands traded with islands as far as the Marianas, and some of them settled there after 1818. The Yapese canoes were also known as flying proas, and among the best in Micronesia. A German trading station was established in 1869. The copra trader O'Keefe became famous for transporting stone money from Palau to Yap in return for copra. Spain claimed the islands in 1874, and they were sold to Germany in 1899. The Japanese came in 1914, followed by the Americans during WWII. Yap became a US Trust Territory.

Top: Aerial view of the main islands of Yap proper, consisting of Rumung, Maap, Gagil/Tomil, and Yap.

Center: Traditional stone money discs. Most of these heavy discs were quarried on Palau, you can read more about this in the Visitors chapter.

Right: The Yap channels are known for the many beautiful Manta rays, which, aside from the culture, are Yap's main tourist attraction.

Shell money

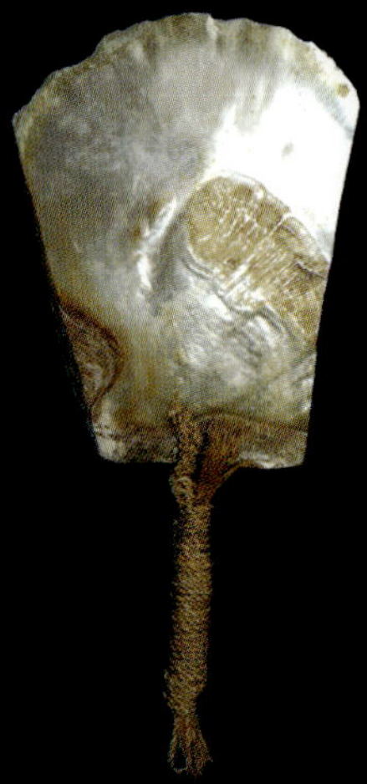

Aside from the stone money, Yapese traditionally use shell money. In the past, mat bundles, banana fiber, clamshell pestles and turmeric were also used as money.

Yar is money made from oysters, the big ones are cut on the sides, and have a woven handle. Smaller ones are strung on coconut rope cords, and used for smaller purchases. The larger oysters were brought in from Palau, Indonesia, and the Philippines. This money is still used today as payment for weddings , celebrations, and apologies. It is only used on Yap proper.

Gau are bead necklaces made from Spondylus shells, with a few Sperm whale teeth. Some necklaces use bone, Dugong, or pig teeth instead. The first of these Gau came from Chuuk centuries ago, and none were made after the early 1900s. These usually 2 meter long necklaces were payments for war, murder, or meeting houses, and the outer islanders were not allowed to own any. Finer made necklaces and belts as shown above were brought from Palau, but are less valuable than the large, rough Gau worn only by Chiefs.

Below: Gau necklaces can be seen during traditional dance performances and on Yap day celebration. The second photo shows the famous local architect John Tamag with his Gau necklace and oyster shell money. The left photo shows a Chief wearing and holding two large Gaus during a village dance performance.

Outer islands

Only the islands of Ulithi, Fais and Woleai are accessible by plane, and many of the outer islands of Yap (except for Ulithi) are still very remote and traditional, and canoes still play an important role. Traditional houses and clothing are still used. Full body tattooing with soot is rare today, but was common in the past. The islands of Ulithi, Satawal and Woleai have their own language, different from Yapese. The people fish, harvest Green turtles and their eggs, and eat sea birds. These islands are regularly hit and devastated by strong typhoons. Special permission is needed to be able to visit some of these islands.

Valuables of the outer islands are lavalava skirts, traditionally woven from banana or hibiscus fiber with intricate clan designs, today made in simple striped design from cotton. Also valuable are turmeric, used for traditional medicine, dancing and spice, as well as woven Betelnut fiber, and handmade coconut fiber rope.

Below: Life on Fais Island is an odd mixture of tradition and modern imported items such as plywood, tin roofs, plastic, and concrete.

Below, left: A woman on Fais Island, weaving a cotton lavalava skirt.

Left: Loincloths are still worn on the outer islands, usually blue, red or white cotton, but also the traditional woven fiber ones for special occasions.

Top: Photo of a Yapese man with full body tattoos, early 1930s.

Coconut rope

Coconut fiber and tree bark fiber have been used for centuries by people of the oceanic islands to make ropes of different sizes to use for tying canoes, sails, fishing traps and lures, and houses. With the import of nylon and cotton rope, the making of coconut rope is a disappearing craft. Only the more remote islands still make the coconut ropes now in different sizes. Very fine and strong hibiscus rope coils, left, are not being made anymore, and only kept by older people as heirloom items.

Left: Traditional fishing lures of the outer islands were made from an oyster shell sinker with a turtleshell hook, and some feathers of a frigate bird tied together with very fine hibiscus and coconut rope. These lures are now replaced by metal fishing hooks and nylon fishing lines. Watertight wooden tackle boxes were used to store the lures on the canoes.

Below: Making fish traps on Woleai Island. These type of traps are placed between corals in shallow water. Most Micronesian islands used similar traps in the past.

Next page: The largest open water traps, over 6 feet long, which are tied to floating logs on the surface in open ocean, are owned by the whole village and hung in the canoe houses when not used.

Palau

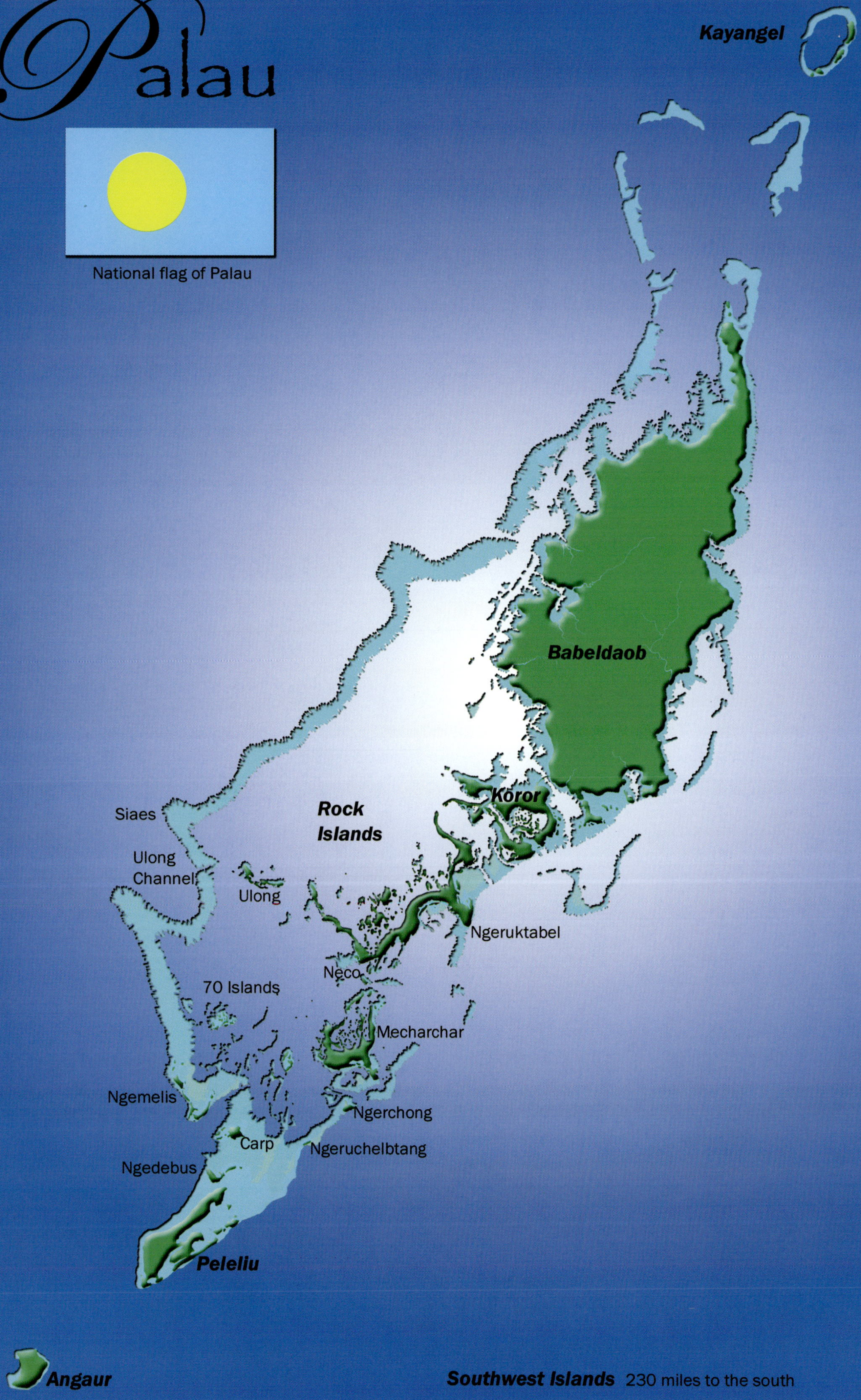

National flag of Palau
Kayangel
Babeldaob
Koror
Rock Islands
Siaes
Ulong Channel
Ulong
Ngeruktabel
Neco
70 Islands
Mecharchar
Ngemelis
Ngerchong
Carp
Ngeruchelbtang
Ngedebus
Peleliu
Angaur
Southwest Islands 230 miles to the south

Political status: Republic, independent since 1994 under a Compact of Free Association with the United States till 2044. It is an American-style democracy with 16 states and 3 branches of government.

Population: Pre-contact estimated at 50,000. Reduced to 5,000 by 1900 due to diseases. Currently 19,219. 13,209 of these are resident Palauans, 5,920 foreign residents (Census 2000). 80% lives on Koror. Another 7,000 Palauans live and work abroad.

Currency: US dollar
Main income sources: US Compact money, tourism. trade.

Location: 550 miles from the Philippines, 813 miles from Guam, 2,500 miles from Japan, 450 miles from Indonesian and Papua New Guinea Islands.

Climate: Mean daily temperature 81 F/ 27 C., with an average humidity of 82%. Average rainfall 150 inches annually.

2003 visitor arrivals: 63,328 **Total hotel rooms**: 1,049
44% Taiwan, 33.8% Japan, 6.7% US, 1.4% Europe.

Islands: About 350 islands, only 12 are inhabited. High volcanic islands (Babeldaob and part of Koror), limestone (Rock Islands, part of Airai, Koror, Peleliu), low coral islands and atolls (Peleliu, Angaur, Southwest Islands, Kayangel). Attractions include the famous Rock Islands with white sandy beaches, snorkeling, scuba diving, kayaking, bird watching and land attractions such as the stone monoliths, WWII sites, and waterfalls.

Below: The current President of Palau Tommy Remengesau Jr. strolling with his daughter.

Schools: There are 18 American-style public elementary schools, and 2 private ones. There is one public high school, 5 private ones, and one vocational college. Palauans have access to US colleges and universities through scholarships, and many graduates have not returned to Palau.

Economic activity: Palau has a GDP per capita of $5,482, which is more than 3 times higher than similar Pacific islands. Imports and hourly wages are higher than most islands also, and so is the cost of living, but tax is still low. There is very high government spending and foreign assistance, and an urbanized population with above average services and employment. Only a very small percentage of the revenue comes from agriculture and inshore fishing.

Challenges: Over time, the Compact grants will be less than the previous level of US funding, and Palau will face a financial crisis, resulting in a lower quality of life and income levels for Palauans. To pursue economic growth, the Palauan government needs to develop a smaller, more efficient government, and expand its private sector and high-end tourism industry. Tourism focus in recent years has changed from high-end scuba divers to low-end tour groups from Taiwan, causing serious environmental concerns, and less overall benefit and revenue.

Health: There is increasing incidence of chronic illnesses such as cancer, diabetes and cardiovascular disease among the Palauans, as well as health problems caused by drugs and alcohol abuse.

Kayangel
Ngeriungs
Ngerebelas
Orak

Kayangel

Kayangel is a classic coral atoll 25 miles north of Babeldaob, and consists of four islands, namely Kayangel (Ngcheangel), Ngeriungs, Ngerebelas, and Orak. They are surrounded by beautiful beaches on the western side, with a calm turquoise lagoon, and rocky wave-beaten shores on the east side. The 130-some people live in one village on the main island of Kayangel, which is the northernmost inhabited island of Palau. The people here were known for their pandanus weaving and for making the traditional grass skirts out of the fiber of hibiscus tree trunks.

A state boat usually goes from Koror to Kayangel and back twice a week, but takes 5-7 hours. Even with the faster but expensive speedboats of the tour operators on Koror the trip takes 2 to 3 hours, and is weather dependent. Check with their state office first about permits and accommodations/camping. There is not much to do on Kayangel, but it is a great place to relax and enjoy fishing, beaches and nature. Diving is good, but the main attractions are above water here. Aside from the far Southwest islands, the three uninhabited islands of Kayangel are the only place on Palau where the large Frigate seabirds roost and nest. These islands also have a large number of Monitor lizards.

The Velasco reef north of Kayangel is a favorite fishing area for Palauans. Velasco is the submerged reef of the atoll of Ngeruangel, which now only has a small sandy islet where terns and an occasional turtle nest. It is believed to have been an atoll with large islands centuries ago, and that it's population moved south after the islands were flooded by an enormous tidal wave.

The island of Kayangel was hit by a strong typhoon in 1992, which damaged the island and surrounding shallow reefs. Local people believe that a sacred hibiscus tree, **Chermall,** *(Hibiscus tiliaceus)* which grows on the southern end of the main island, protects them from rough seas. Today, they still bring offerings to the stone face located next to the tree. *(You can see a photograph of the tree and stone face in the next Pre-contact chapter)*

Babeldaob

Ngardmau waterfalls

Ngerchelong stone monoliths

Ngatpang waterfalls

Aimeliik bai

Melekeok bai

Airai airport

Airai bai

Babeldaob, or Babelthuap, is the second-largest island in Micronesia after Guam, and is mostly volcanic, aside from some limestone areas in Airai state. It is home to ten of the sixteen states of Palau, each with only a few hundred people. Although your airline ticket will say Koror, you actually land at the international airport in Airai state on Babeldaob, and drive over the K-B bridge, now called the new Japan-Palau friendship bridge, to Koror.

The highest point on Babeldaob is Mt. Ngerchelchuus at 713 feet in Ngardmau state. Ngardmau is also known for the largest waterfalls on Palau. Lake Ngerdok in Melekeok state is the largest natural freshwater lake on Palau. Most of the traditional villages throughout Babeldaob are connected by ancient stone pathways, some of the best preserved ones can be seen in Ngaraard state, while the dense interior forest of Babeldaob is mostly unexplored. Ancient stone monoliths and mysterious terraces are found throughout Babeldaob, the largest monoliths are found in Ngerchelong state. Airai has the largest population (around 1,400) because of it's close proximity to Koror. It also has the only remaining original Bai meeting house, which was constructed in the late 1800s. Recently built Bais can be seen in Melekeok and Aimeliik.

Melekeok state, which was one of the most powerful districts before foreign contact, was designated to become Palau's future capital. The new capital building, below, is currently under construction there, financed by the United States and by a grant from Taiwan.

Since 1999 the construction of a paved $150 million dollar 53 mile road is also underway, by a Korean contractor and financed by the United States under the Compact of Free Association. The clay soil, erosion problems and much more rain than expected in recent years have made this project difficult and way behind schedule. The erosion and run-off from the road project is visible around Babeldaob Island from the air, and may have an enormous impact on the mangrove areas which line the western coasts and which provide an important run-off filter for the coral reefs.

Right: Before, the dirt roads on Babeldaob were a real challenge during rainy days, and cars regularly got stuck. After 5 years of construction, the new Compact road is now much improved, but far from completed.
Below: The new capital building in Melekeok state is nearing completion, but not many local people are happy with the grandiose and unpractical design.

the KB bridge disaster

The 790 feet long Koror-Babeldaob bridge, or KB bridge as it was known locally, was constructed during 1975-1977, and cost $5.2 million dollars. Designed to withstand typhoons and earthquakes, it was the longest bridge of it's type in the world, with an anticipated life-span of 20 years. The two sections, hinged together in the center, were constructed out of concrete, reinforced steel and steel cables, and they were able to stand on their own.

In the late 1980s, the bridge had already visibly started sagging in the center, and it became worse every year. We used to joke that sooner or later the bridge would collapse, but it was still a shock when it actually did collapse, my son and I witnessed the disaster from our house. Major repairs to the bridge had just been done a few months earlier at a cost of $2 million dollars. Lawsuits regarding the repairs were later settled out of court.

The original bridge. All the traffic from the Babeldaob and the international airport, as well as power, telephone and water depended on this bridge to get to the island of Koror.

The bridge collapsed on September 26, 1996, at 5.30 PM. It left 2 people dead and 4 injured. It could have been a lot worse, considering the fact that 5 days later hundreds of Palauans would have stood on this bridge to watch the October 1st Independence Day boat races.

Cables were hung across for much needed water and power in Koror. This was restored within weeks, and transportation across the channel by boat was available the same day the bridge collapsed.

A temporary floating bridge, good for 5 years, was installed while the work on removing the remains of the collapsed bridge started. A grant from the Japan government paid for Palau's new bridge.

The new Japan-Palau Friendship bridge was officially donated by the government of Japan in December 2001, five years after the old bridge collapsed. The new bridge is a 3-span hybrid extra-closed bridge type, over 1,350 feet long. The two opposite sides were constructed toward the center, and the large center piece was installed last, lifted up from a barge.

Below: The first Palau Independence Day celebrations on the new bridge, on October 1st, 2003. Boat races took place around the new bridge, with hundreds of local spectators watching from around the bridge and on boats.

Aerial view of Koror and the surrounding Rock Islands.

Malakal Island, below, is missing from the above photo, but would be on the left. The main harbor, and most of the tour and dive operators are located here, as well as some smaller hotels, and the clamfarm.

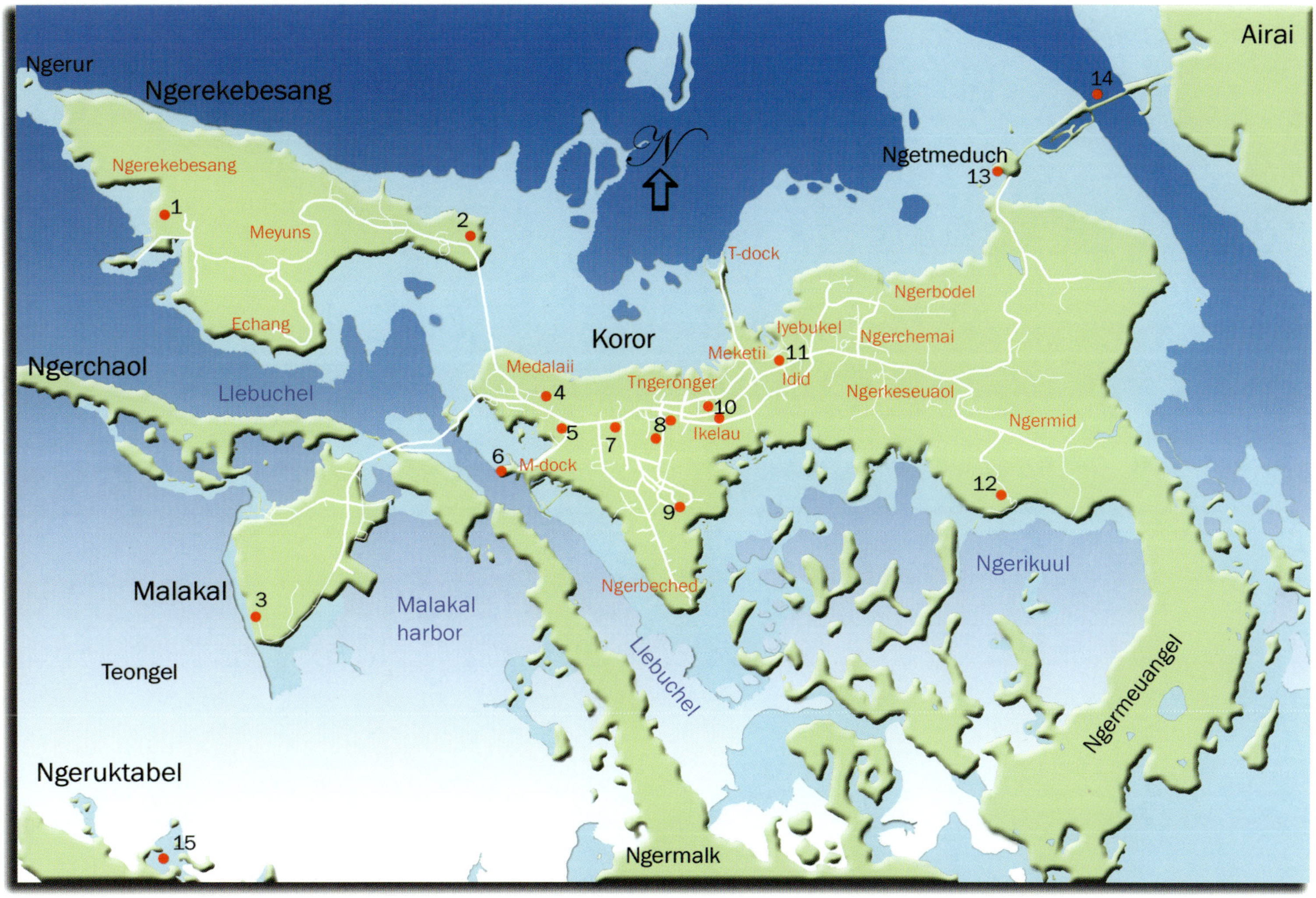

Village names in red Landmarks:

1. Palau Pacific Resort
2. National Hospital
3. Palau Mariculture Center
4. Cultural Center
5. Courthouse
6. Palau International Coral Reef Center
7. Bank of Hawaii and Neco Plaza buildings
8. Post office and Police Station
9. Belau National Museum/ Bai
10. Ben Franklin/Surangels shopping centers
11. Etpison Museum
12. Stone Face
13. Crocodile Farm
14. K-B Bridge.
15. Dolphins Pacific

Koror State consists of the islands of Koror (Oreor), Malakal, and Ngerekebesang (Arakebesang) which are connected to each other by causeways, and also all the uninhabited Rock Islands north of Ngedebus and Carp Island. Koror itself is the economic center and current capital of Palau. Most of the tour operators and hotels are located on Koror, and there are excellent restaurants down town. There are no beaches on Koror aside from the man-made beach at the Palau Pacific Resort. To get to the best beaches and snorkeling you have to take a boat to eastern Babeldaob, or down to the Rock Islands.

The majority of Palauans live and work on Koror, and most of the sixteen states have an office on Koror. Almost every family on Koror has one or two cars, and traffic on the one main road during weekdays is bumper to bumper every morning between 7.30- 8.30 AM, and in the afternoon between 4.30- 5.30 PM. Biking and walking can be tricky, especially at night, due to the lack of sidewalks, aggressive dogs, and bad or drunk drivers. Travellers checks and credit cards are accepted in most hotels and stores on Koror, but you have to have cash dollars with you when staying on Babeldaob, Peleliu or Angaur. There are three private out-patient clinics on Koror, and the Belau National Hospital is located nearby on Ngerekebesang.

Attractions on Koror include the PICRC aquarium; the Etpison Museum which has an outdoors mural by the artist Wyland showing the endangered Dugong seacow: the new Belau National Museum which has a traditional Bai next to it: the new Cultural Center: and Dolphins Pacific, where you can interact and swim with dolphins, is only a 5-minute boat ride from Koror. There are stores with handicrafts and storyboards, and the two largest stores, Ben Franklin and Surangel's, both have groceries downstairs and department stores upstairs.

During the Japanese occupation in the 1930s, Koror was a busy town with over 30,000 Japanese, paved streets, shrines, stores, and later military facilities. After being bombed flat by American planes in 1944, Koror was slowly built up again, and some remnants of the Japanese time can still be seen today, like the courthouse.

The Palau Pacific Resort property, built on top of a Japanese WWII seaplane ramp, has been the best hotel on Palau since 1984.

The Malakal Lighthouse channel.

Rock islands

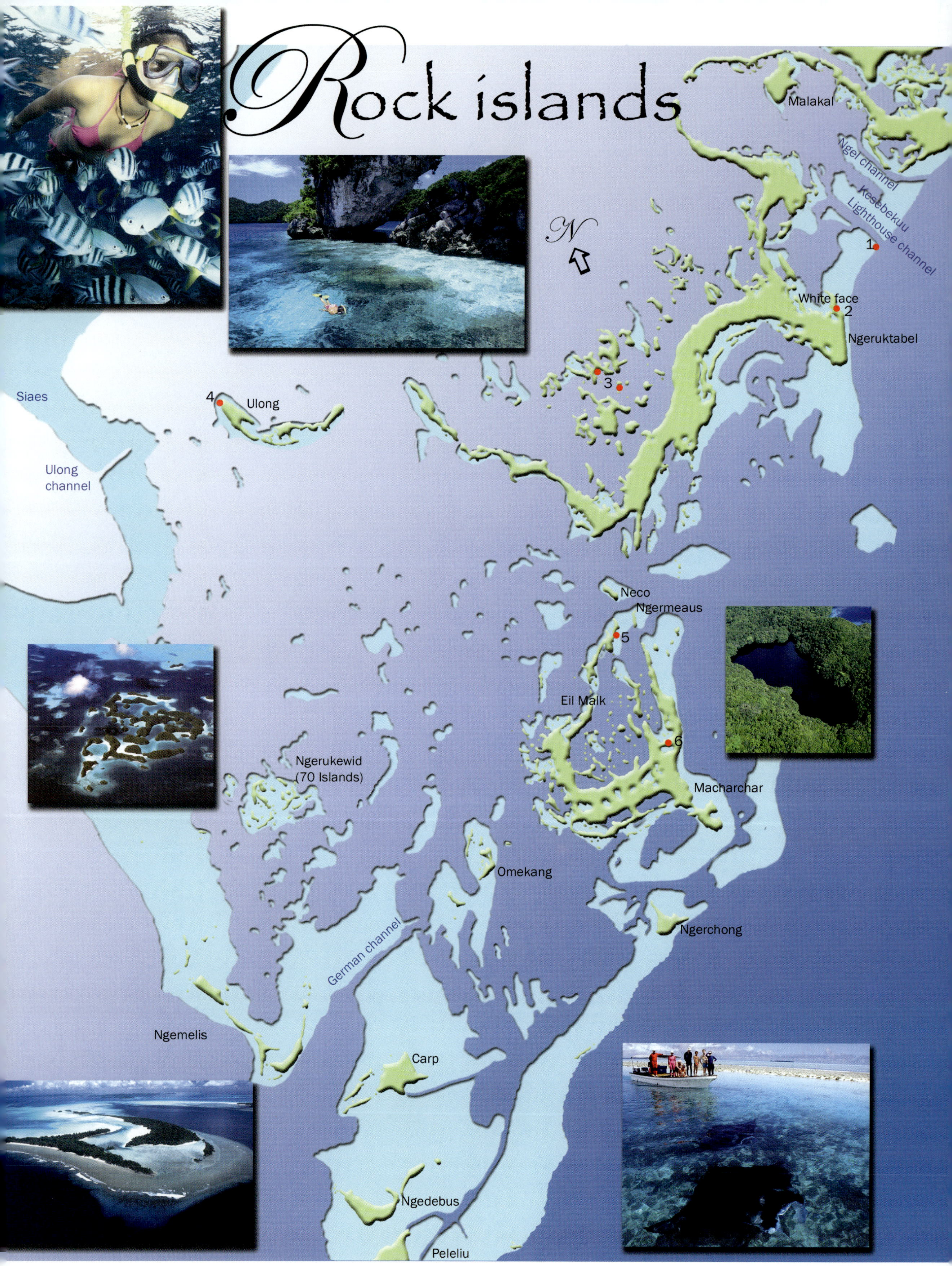

If you are lucky, the pilot of your airplane will take a short detour over the Rock Islands before landing, to give you a glimpse of the mushroom-shaped green islands, white beaches, and turquoise colored lagoons and reefs. The Rock Islands have the best snorkeling, beaches and nature Palau has to offer. They are uninhabited, although most of the larger public beaches now have shelters and toilets for tourists, and Palauans often come to spend the weekends to fish and hunt. The limestone islands are covered in dense green jungle growth, and are still full of birds, bats, and crocodiles. Hiking and exploring beyond the beaches can be great fun, as long as you bring insect repellent, long pants and good shoes. Ask someone to show you what the local poison tree looks like so you can avoid touching it, and watch out for thorned vines.

Ancient rock paintings, burial grounds, pottery, and other evidence of early inhabitants can still be found, but by the time of the first European contact in 1783 the Rock Islands had been uninhabited for quite some time. Yapese quarried their limestone disc money here, and some of the caves close to the waterline still have Japanese artillery and WWII artifacts. Japanese ship and plane wrecks are scattered around the Rock Island lagoon, shot down by American planes in 1944. A few of the American planes that were shot down can also be seen. The unique Jellyfish Lake is a must for visitors to see, and there is great snorkeling throughout the lagoon, while the outer reefs, like the dropoffs of Ngemelis, offer some of the world's best scuba diving and marine life action.

Blue Corner is one dive spot that has made Palau famous among divers, with its strong currents attracting reef sharks and huge schools of fish. The nearby Blue Holes are also spectacular, and so are about a dozen other dive spots in this area. Please note that it is illegal to collect live shells from the Rock Islands, and to fish at the dive spots.
The picture-perfect 70 Islands cannot be visited, they were set aside as a wildlife preserve in 1956 to protect nesting turtles and birds, but equally beautiful islands are found nearby. German channel is where all boats cross over the reef. Manta rays often pass through this channel also, and just outside the mouth of the channel is a Manta cleaning station, where divers can watch these gentle giants getting cleaned by small fishes. Check with local tour operators on the latest permits and regulations before you arrive.

Map details previous page:
1. Lighthouse
2. White Face ancient rock paintings
3. Soft Coral Arch & Bridge Island
4. Ulong ancient rock paintings
5. Clam city
6. Jellyfish Lake

The Ngerukewid (70 Islands) preserve is off-limits to visitors and can only be enjoyed from the air.

Pelelíu

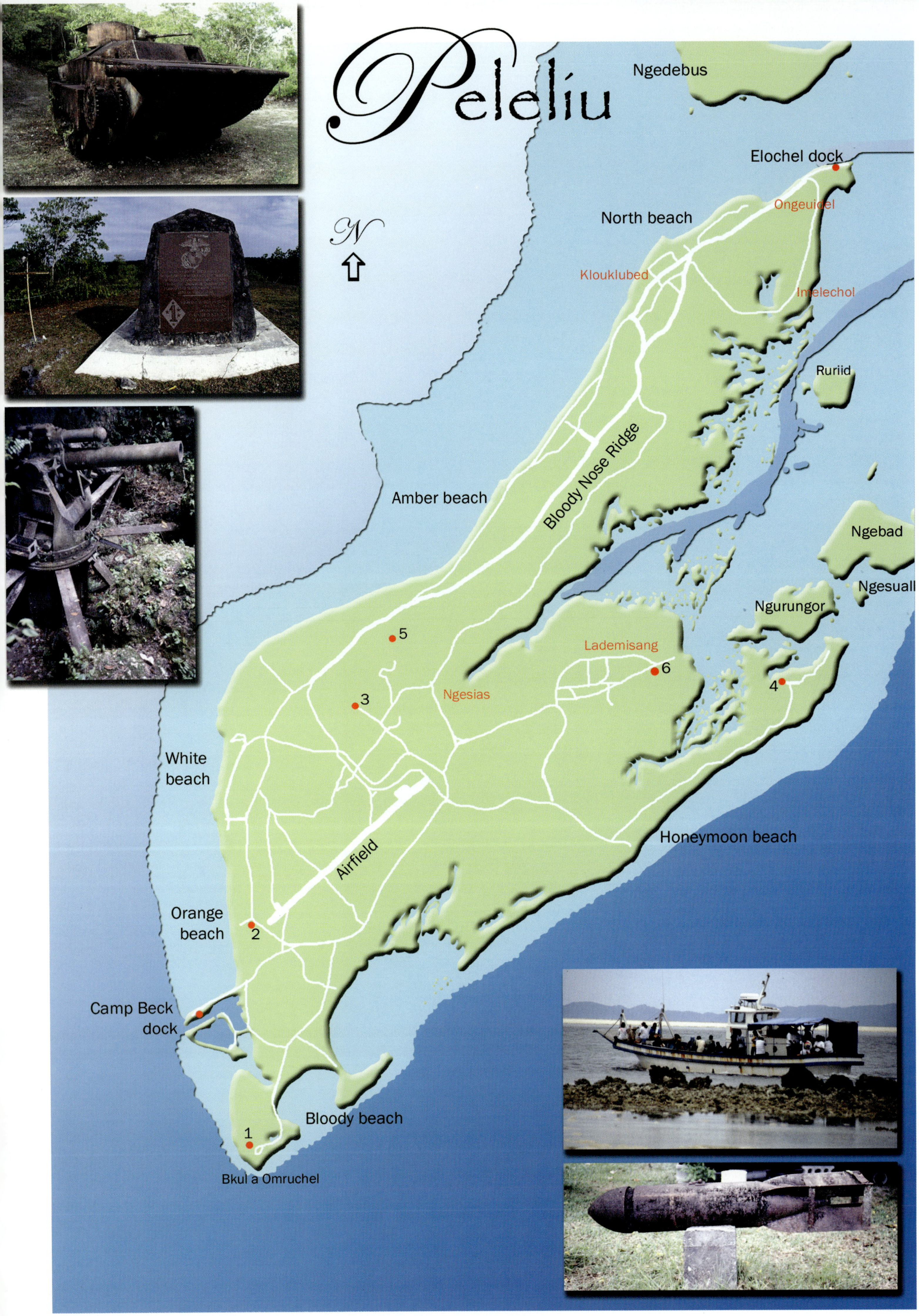
Ngedebus
Elochel dock
North beach
Ongeuidel
Klouklubed
Imelechol
N
Ruriid
Bloody Nose Ridge
Amber beach
Ngebad
Ngesuall
Ngurungor
Lademisang
5
6
4
3
Ngesias
White beach
Airfield
Honeymoon beach
Orange beach
2
Camp Beck dock
Bloody beach
1
Bkul a Omruchel

Peleliu is locally known as Beliliou, and was also known as Odesangel (The place where it all began) in the past. Palauans believe most settlers of northern Palau originally came from Peleliu. Before Koror became more powerful with the help of foreigners, Peleliu was an important district in Palau.

The Japanese started fortifying the island in 1944, and forced the locals to move to Babeldaob. Away from their villages and crops, they had to survive in the jungles and mangroves. In September, 1944, after heavy aerial bombing, the Americans landed on Peleliu during Operation Stalemate, an unnecessary and costly battle. Over 10,000 Japanese were defending the island to death from limestone caves in the center of the island. Peleliu and nearby Ngedebus were finally captured by the Americans after 2 1/2 months of fighting, with over 1,800 Americans, and over 10,000 Japanese killed. Only about 400 prisoners were taken, mostly Korean laborers. There were 5 Japanese soldiers who survived inside the Peleliu caves for a year, when they were induced to come out. The island today is still littered with live ammunition and mines, so do not wander around without a guide. You can visit American and Japanese memorials and caves, and WWII artifacts such as rusty tanks and heavy artillery are still sitting around. It is illegal to remove any WWII artifacts from the island.

Peleliu has several excellent dive spots, including Peleliu Corner and Peleliu Express, which are dives for advanced divers in small groups only, with an experienced guide and boatdriver. Currents from several directions come together at the southern tip of this island and can get extreme, quite a few people over the years have ended up floating out to open ocean here, usually because they got caught in the current and surfaced too far away from the boat. There is a lot of fish and shark action because of the currents, silvertips and thresher sharks are often seen here. In the winter months Orcas and Whale sharks have been spotted off the southern reef tip.

Map details previous page:

1. Peleliu Peace Memorial Park
2. 81st Infantry Division Memorial
3. Japanese cannon and tank
4. Ngermelt swimming hole
5. Bloody Nose Ridge monuments
6. Peleliu shrine

Angaur

Blowholes
Buddhist Memorial
Santa Maria statue
Nguruhei beach
Phosphate ponds
Caves
WWII airplane
Lighthouse
Rois
Angaur harbor
Airfield
Ngaramasch
Ngedloch beach
N

With less than 200 people, Angaur Island is nevertheless one of the sixteen states of Palau. Seven miles south of Peleliu, Angaur is outside the protective reef that surrounds most of the rest of Palau. The northern coast is usually rough, with a rocky coast and dramatic blowholes, while the southern side is much calmer, and has beautiful sandy beaches, and a sheltered harbor. There is a state boat that goes back and forth several times a week, and you can also charter a boat, or fly down to Angaur from Koror. There are several guest houses available with simple meals. Angaur is very laid-back, and great for nature lovers and photographers. The best way to get around is by bike.

The beaches of Angaur have a more orange colored sand than elsewhere on Palau. Ngedloch beach is known as the place where the spirits of all dead Palauans stop before passing over to the spirit world. In 1944, the Japanese, who had been mining for phosphate on Angaur, brought in 1,400 troops and fortified the island. The Americans landed in September, 1944 during operation Stalemate, and took the island after 4 days of fighting.

Hundreds of Crab-eating macaque monkeys inhabit the dense forest, of Angaur, accidentally released during the German time (1899- 1914). These smart animals have adapted to life on Angaur without any problem. They eat almost anything, including crabs, bird eggs, insects and fruits, and are good swimmers, even holding their breath under water. The locals have always considered the monkeys a pest, because they would steal fruits and vegetables from around their houses. The population of monkeys seems to have increased a lot lately, and the final straw for the Angaur people was when the monkeys started eating and taking the locally grown betelnuts. There are now attempts underway to eradicate the monkeys and stop them from spreading to the other islands. The babies are popular pets, and often brought to Koror. Today, escaped and released monkeys have already started living in the wild on Koror, and pose a threat to endemic birds and other native species.

If you go around the island, you are also likely to see large Monitor lizards, and land crabs. The freshwater lakes, previously used as phosphate mines by the Germans and Japanese, are now home to birds and crocodiles. There are several WWII plane wrecks on land, including a F-4 Corsair fighter and B-29 bomber.

Diving around Angaur is very good, with amazing visibility. The reefs have a lot of lobsters and shells, and fish life is spectacular, sometimes you can have encounters with huge schools of Bumphead parrotfish, tuna, or large turtles and pelagic sharks like the Oceanic white-tip. Conditions can be rough, and since it is quite far from Koror, none of the dive operators go there regularly. The destroyer USS Perry struck a Japanese mine in September, 1944, and the ship was recently found off Angaur in 260 feet of water.

Sonsorol & Fana

Pulo anna

Merir

Tobi

Helen reef

Southwest islands

The remote Southwest Islands consist of six islands, the largest is Sonsorol (Dangosaro), then Merir (Meilieli), Tobi, Fana (Fanna), and Pulo Anna (Purr). Helen Reef (Hotsarihei) is the smallest, but is the only atoll, with a larger reef formation than any of the other islands. The closest islands are Angaur to the northwest, Morotai (Indonesia) to the west of Tobi, and Kepulauan Asia (Indonesia) to the south of Helen reef, all at least 240 km away.

Political status: These islands are divided into two states under the Republic of Palau, The State of Sonsorol (Sonsorol, Fana, Merir, Pulo Anna) and Hatohobei State (Tobi and Helen Reef). Each has a governor and state office on Koror.

Population: Estimated at over 1,000 before 1905. Only few people now live on these islands. Fana is not inhabited, while about 40 people stay on Sonsorol, and another 30 on Pulo Anna. Merir has only 2 people, Tobi only has about ten people at the moment, while only three are staying on Helen Reef. Most people from the Southwest Islands stay on Koror, in the village of Echang, while others go back and forth for medical treatment or schooling. The people eat fish and chicken, or hunt for coconut crabs, turtles and their eggs, and seabirds. Dogs brought in by the people are a threat to the nesting seabirds, most of which have moved to Fana.

History: These islands were last settled by people from the outer islands of Yap, and they speak a different languages, similar to Yapese. Archaeological research indicates the islands were already inhabited long before that. Canoes from the Caroline Islands occasionally floated to the Philippines, sparking the interest of Spanish priests. After finally spotting Sonsorol, two priests of a Spanish ship were left there in 1710, and unable to be picked up due to bad weather, were never heard off again. In 1832 the American whaler Mentor wrecked off Palau, and some of the crew, trying to sail to Indonesia, ended up as prisoners on Tobi Island, which they called Lord North's Island. (see Visitors chapter). After a major typhoon struck the islands around 1905, and the surviving people came to Koror and were given a piece of land to live on. The Japanese occupied the islands until the end of WWII, and mined for phosphate on Sonsorol and Tobi.

Attractions: Green turtles nest in large numbers on Merir and Helen Reef, and large seabird populations of Boobies, Noddies and Frigate birds nest on Helen Reef and Fana. The best time of year to visit is between April and July, when the weather is relatively calm and the turtles and birds are mating and nesting. No accommodations are available, and you have to bring your own supplies and food. The government boat visits infrequently, and occasional trips are organized by live-aboard vessels operating out of Koror. Anchoring is tricky on all islands except Helen reef, and the weather can change overnight. You do not want to stay overnight on Merir, which has a mangrove swamp in the center of the island, and is infested with mosquitoes. The diving is good, with clear water, undamaged reefs, and occasional whale sightings, but the fish life is not abundant. For fishing and nature photography these islands are great. The few people there do not live in the traditional way anymore, so do not be surprised by the tin roofs and T-shirts.

Supply boat arrives at Tobi Island.

A Red-footed Booby chick looks down from its nest on Fana Island.

A Green turtle laying eggs on a secluded beach at night.

Pre-contact

The first settlers of Palau may have arrived after accidental drift voyages. Even in recent years several boats and canoes that were lost at sea have floated into Palauan waters from Indonesia, Melanesia, and the southern Philippines. Archeological research shows Babeldaob and the Rock Islands were inhabited as early as 1000 BC, and the population before 1783 has been estimated at 40,000- 60,000. Oral history suggests the first settlers landed on Angaur, Peleliu and the Rock Islands before moving north to Babeldaob. Ceramic bowls painted with red designs have been found in two limestone caves, and dated to 640 BC. These are very similar to excavated Neolithic Chinese pottery.

Originally the Palauans were probably accomplished sailors and navigators, but by the 1700s, their canoes were built for domestic inter-island travel only, and fishing was confined to the lagoons and nearby reefs, except for the island of Angaur, which had oceangoing canoes. Unlike other Micronesian atolls, whose survival depended on their open ocean voyages, the northern Palauans had all the natural resources they needed on their own land and within their reefs. One type of the Palauan money beads that have been in use on Palau for centuries, were manufactured in Indonesia between 600 and 900 AD, suggesting some contact with Indonesia or traders during that time.

Primitive art is found high on the limestone cliffs of the Rock Islands which have galleries of red and yellow rock paintings of unknown age, probably dating back at least a thousand years . The paintings show simple handprints, human figures and canoes, as well as complicated line and geometrical designs, but none show any resemblance to the later Bai paintings. The largest rock painting of a human or spirit is 3 feet high, and has a face strangely similar to the large stone carved faces found on Babeldaob, with hollow eyes and a fanged mouth. Also fascinating is the variety in the carving style of the dozens of stone carved faces scattered around Babeldaob, also of unknown age. Some are very finely carved with a lot of detail and smiling, while others are roughly made and obviously meant to intimidate.

Judging from pottery sherds dating, the Rock Islands were again inhabited between 1200 and 1400 AD, then abandoned. The terraces on Babeldaob had most activity between 1000 and 1400 AD.

Clamshell handtool, found in limestone burial cave.

Trochus shell bracelets, found in a limestone burial cave.

Oyster shell knives, still used today by Palauan women, were often buried with the dead.

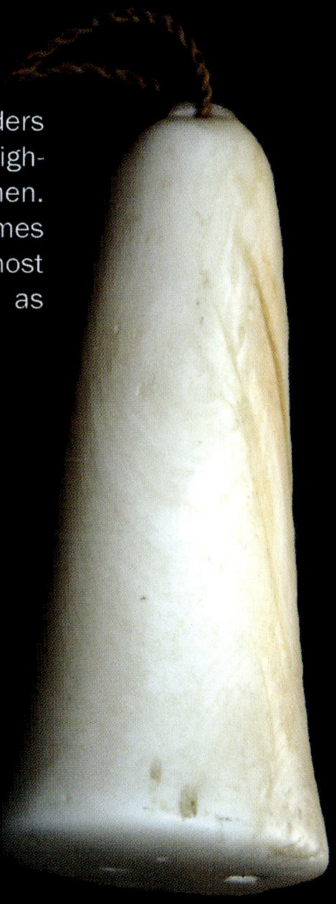

Clamshell pounders were used by high-clan Palauan women. They are sometimes excavated, but most have been kept as heirloom items.

The adze was the favorite men's tool, used to carve canoes, houses, etc. Before metal became common in the late 1800s, the adze blades were made from sharp Terebra shells or Tridacna clamshell. These shell blades are excavated all over Palau.
Bottom: An excavated small adze blade, made from the tooth of a Dugong seacow.

Other similarities are found between the outer islands of Chuuk, Yap, Palau, and the Trobiand islands east of New Guinea. The finer type of shell Gau necklaces, believed to have been made on Chuuk and Palau in the past, are still made today and used as valuable necklaces as they are on Yap and were on Chuuk in the past. The Trobiand women wear woven waistbelts as the Palauan women used to do, and after the death of a chief, the women carry around a box with his personal items for several months. It is said the box contains his spirit, and touching it will make you feel him. This is also done on Palau in the same way. Egg cowries play an important role on the Trobiand islands and are used for decorating houses, canoes, and as jewelry. Canoes on Yap and Palau were decorated only with egg cowries in the past, but the meaning for this has been lost. The Trobiand islands also have painted spirit houses similar to the ones that were found on Palau in previous centuries.

There are records of Yapese people settling several villages on Palau, people from Ulithi settling the Southwest Islands, and people from Makassar, Indonesia, apparently settled a village on Babeldaob. (Kubary, 1885) By the time the first Europeans arrived in 1783, Palauans had never seen white men before, but were familiar with Asian traders and other Pacific islanders.

Coconuts probably floated in from Southeast Asia, while certain plants were probably brought in by birds through seeds in their stomach contents, or pollen between their feathers. Seeds could also have arrived with the wind or on floating driftwood. The saltwater crocodile may have floated or swam from New Guinea. An excavation find showed that pigs were present on the Rock Islands of Palau from 500- 900 AD, long before the Europeans arrived.

Undoubtedly there are other Pacific islands that have striking similarities with Palau, where some of the first settlers may have come from, but most likely there was a mix of settlers from Indonesia, Malaysia, Melanesia and the Philippines over the centuries, as well as some inter-island migrations within Micronesia. Hopefully more studies will be done on some of the specific cultural traditions similar between all these islands before they will totally fade away in the new Americanized way of life most of the islands are adopting today.

What we know for sure, is that before foreign contact, the Palauans had a well-established complex society based on a matrilineal clan system. Each district had villages with their own council of chiefs and clan structures. Villages formed alliances, exchanged services and goods, and had fierce competition and warfare between them. The two areas that became the most powerful were Melekeok on Babeldaob, and Koror. Their highest chiefs, Reklai and Ibedul, became the paramount chiefs of the whole Palau. Female chiefs were responsible for selecting the male chiefs from a certain clan. Women were often exchanged between villages, voluntary in exchange for moneybeads, but also as the spoils of war.

Below: Volcanic stones with faded red crosses painted on them were placed next to dead peoples' skulls in one of the limestone burial caves found.

This badly eroded coral sandstone figurine shows a seated male grasping his now broken off penis. This sexually explicit carving was collected by a Peace Corp volunteer from Tobi Island in 1968. It is believed to be one of a set of two that may have been used in sex education in the past, and may be ancient (Osborne, 1979)

Monkey men, also called Sen, are carved out of wood by Tobi men, and today sold as souvenirs on Koror. In the past, these statues were used on some Yapese islands as a guardian spirit for the dead. Before the Tobi people moved to Koror in the early 1900s, these carvings used to be placed in the men's meeting houses on Tobi Island.

Rock paintings

Ancient rock art is found on at least 8 different limestone cliff sites on the Rock Islands. The two largest known areas are White Face, shown above, located close to the Malakal lighthouse, and the other area is on the island of Ulong. All of the art is about 40 feet above the waterline, on what seem like natural pathways or galleries cut out of the limestone. Both the White Face and Ulong areas have caves nearby with pottery sherds, but the two sites are very different in painting style and design.

Nothing is known about the origin of these paintings, other than a vague local legend about a God practising his art on the rocks before making the painted Bai meeting houses up north. There are no obvious similarities between the rock art and the Bai decorations in style or meaning, and the rock art was probably done much earlier, and perhaps by a different group of settlers than the later Palauans.

Man in a canoe with enormous hands.

This may represent a canoe on top.

This 3 feet high painting of a spirit or man has a canoe on his head and enormous genitals. The face looks like the stone carved faces found up north.

Ulong Island also has a beautiful gallery of ancient rock art. It is very different from the White Face site. At White Face there are many hand prints, canoes and human stick figures, while at Ulong the only human form is the one shown at left, one of the few paintings where the yellow pigment is still visible. This is one of my favorites, it could be a female form with the yellow being a skirt, or perhaps a snake.

Only Ulong has paintings with yellow pigment, all the other rock art sites found have only red pigment visible. Ulong has mostly complex geometric line designs, which are difficult to explain. Nothing is known about the age of these drawings.

Above: You can see the "Gallery" on the west side of Ulong island from the boat, but it takes a steep hike to get up on the limestone walkway for a closer look. White Face and the other sites are more difficult to climb, and should not be attempted without a guide and safety ropes.

Next page; Some of the Ulong designs. Most of them look almost like stencils, while others are rough and obviously hand painted.

When you continue hiking north passed the gallery of rock ar at Ulong Island, you will encounter several caves. One of these caves has a cooking pit (below, left), and contains the only rock art on Palau ever found deep inside a cave. One painting which looks like an arrow, left, is located in a little niche and easily overlooked, while the other symbol, left middle, is located on a wall behind the cooking pit.

This may have been where the people who made the rock ar mixed their pigments, or lived. Pottery sherds were also found inside and near this cave. Probably red and yellow clay was used for the paintings, sealed with the boiled oil of the Cheritem nut. This was also used to paint and waterproo wooden bowls and canoes in Palau up to the early 1900s.

All the known rock art sites in the Rock Islands were found on exposed cliffs. Judging from the many overgrown cliffs throughout the islands, there may be many more ancient caves and rock art galleries waiting to be discovered behind the dense growth.

Rock art painted in black and white pigments is known from one cave site on Guam (see page 23), and one cave on Rota.

The rock art shown on this page is from the six other sites found scattered around the Rock Islands. The one above, right is from Long Island, on Malakal. Some are badly eroded and faded because they have been exposed to the elements. The one below, left seems to represent a canoe. It was the only painting found on a large exposed cliff.

Next page: This is one of those "don't try this at home" type photographs of my husband Shallum with another Palauan exploring limestone cliffs for rock art. Sharp and loose rocks with little to hold onto make these limestone cliffs difficult and dangerous to climb. Not much of a climber myself, I usually sit on the boat, and only let myself be dragged up onto these cliffs after they find something to take pictures off.

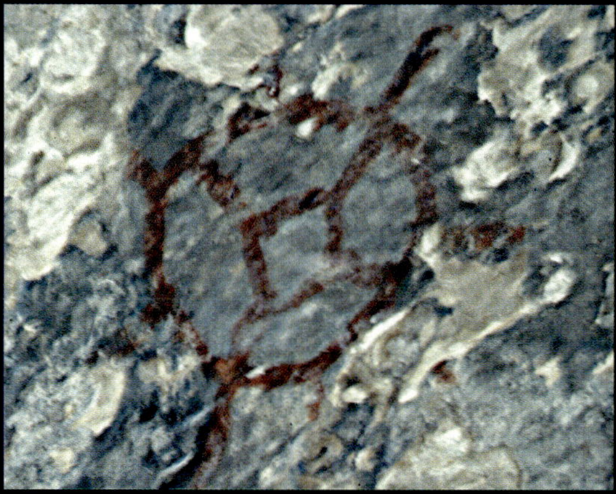

Pottery

Pottery sherds can be found all over Palau, especially in the Rock Islands, even though the clay to make the pots is only found on volcanic Babeldaob Island.

Six ceramic bowls painted with lined red designs were found near a cave on the Rock Islands in 1929 by Hisakatsu Hidikata (see page 73,74). These may have been made by the same people who did the rock art on the limestone cliffs, and are probably the oldest type of pottery found on Palau.

In 1994, my husband found a cave near our house, which is built on a limestone ridge on the east side of Koror, technically part of the Rock Islands. This ridge is believed to be where dead people pass by, and a lot of local people believe the area is haunted. The cave is near a concrete water tank built by the Japanese in the 1930s. The top area of the cave was occupied by Japanese or used as a resting place, and remains of an old lantern, metal and wood were found. At the back of the cave, a hole in the floor is the entrance to a much lower cave, in which we found ancient human remains, a clamshell hand tool, and the ceramic bowls and fragments you see on these two pages. They have the same red painted designs as the bowls found by Hidikata in 1929, and were likely made by the same people.

Two of the four bowls were completely intact, but one, left, was placed under an overhang with water dripping in it, and the interior design and clay was badly eroded. The design on the back of the pot was still very clear, below left.

Probably the same method that was later used for canoes and wood was used to paint these ceramic vessels, locally called **Olkang**. The design would be painted with red clay, and sealed/ waterproofed with the boiled oil of the Panarium nut **Cheritem,** this oil was called **Laok**.

Next page, top: Three views of the second complete bowl found, with interior and exterior designs. The third and largest bowl, bottom right, was in pieces and incomplete. The other piece is one large fragment of a fourth bowl. Due to the dangerously loose bottom structure of the cave, which has several little tunnels and holes and collapses down into a 200 feet high cave with natural stream, we were not able to search or excavate further. The bowls are on display at the Etpison Museum on Koror. One sherd was sent off to the States and dated to 640 BC.

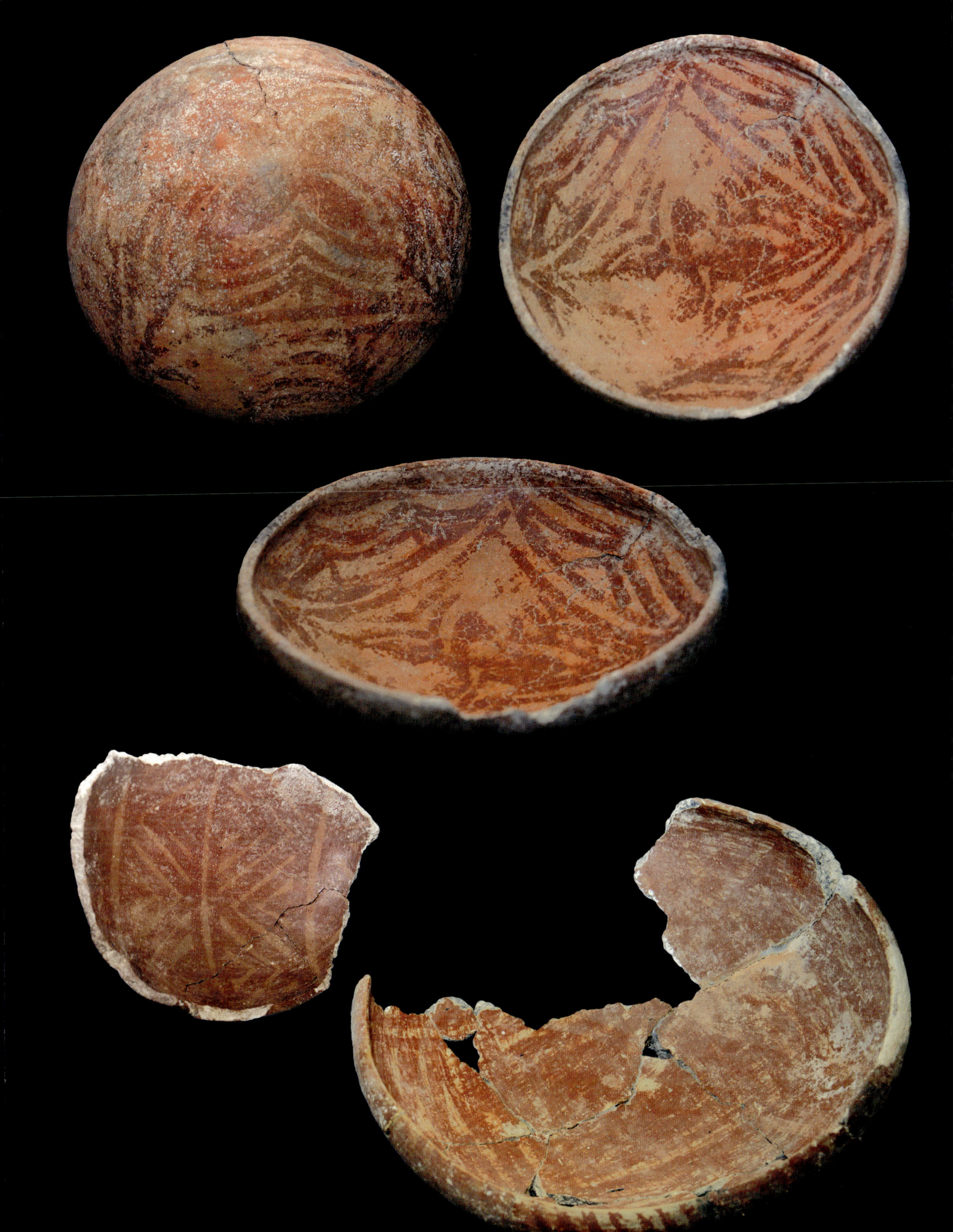

Decorated pottery sherds, above, are found, but rare, and no complete pots with elaborate decorations other than the painted red designs and the pots on the next pages have been found so far.

The decorated bowls shown below and on the next page are from the Belau National Museum collection, and were collected in 1929 by Hisakatsu Hidikata. A Palauan woman found them for him near a cave on Ngeruktabel island, not far from where the White Face rock art is located. There were six of these bowls found originally, but only two survive today. (Hidikata, 1960). The bowl below is only decorated on the outside, but it is the only painted bowl found in this shape and with handles.

The early pottery on Palau was characterized by fine pastes and thin vessel walls, while the later pots are coarser, with thick walls. By the early 1900s, the craft was practically non-existent. In 1954 researcher Osborne had two Palauan women make pottery the way they remembered it from their youth to see the technique and materials used. A coiling method was used and the temper consisted of ground pottery sherds kneaded into the clay.

Clay was obtained from Babeldaob. The best areas for clay were Oikull (Airai), Ngatpang, Ngardmau, and Oll (Ngaraard). Oikull, Ngatpang and Ngardmau were known for making most of the pottery. Most of the pots were made for cooking, called **Olkang**, and some as water storage jars, called **Bekai**.

During the Japanese time (1914-1945) a lot of ceramics were imported. These were stronger because of better firing, and glazed, unlike the local pottery. Locally made pottery at that time consisted mostly of large water pots like number 6 on the next page.

Next page: Number 5 was excavated by someone from Airai, who found this thick complete vessel in his garden in Airai. Number 1, 4 and 7 are thin-walled pots and were found by hunters high on top of a limestone ridge, where they said there were also remains of old stone pathways. Number 2 and 3 are also thin-walled, finely made pots, and were found together with the pot shown below, bones and shell artifacts, in a limestone burial cave. Number 3 is a thin hanging pot, with 4 holes drilled on top.

This complete ceramic pot is decorated with marks along the rim of the pot. Most of the complete vessels found came from limestone caves in the Rock Islands.

1
2
3
4
5
6
7

The first and only petroglyphs ever found on Palau were discovered by archeologist Tim Rieth, who was working along the Compact Road project in Ngaraard State, Babeldaob. The single rock has carvings representing female genitalia. German researcher Augustin Kramer recorded stories from all the Bai meeting houses that were still standing in the early 1900s, and he recorded the story shown above, about the making of the sexes by a brother and sister. This obviously has a connection with the stone, and the carvings are very similar to the drawings from the Bai. The Bai was located in Ngril (Ngerchelong).

Next page: Stone pathways and platforms can be found throughout Palau. Remains of stone pathways have even been found high on top of the Rock Islands. Most of the traditional villages on Babeldaob are connected by stone roads which are centuries old.

Stone artifacts are scattered throughout Babeldaob. Some objects are still used by Palauans to worship and make offerings to the Spirits, like some of the stone carved faces and stone bowls, while other objects sit long forgotten in the dense jungles. The Palau government Bureau of Arts and Culture has been going from state to state in recent years to make inventories of all the historical sites, and to encourage the states to take care of them. Some of the states are now in the process of restoring and clearing their sites, to make them more accessable to tourists.

Left: A stone carving in the shape of a traditional Spirit house is found on Kayangel, on a stone platform where once a Bai stood. The small niche on the side was probably used to place small offerings like betelnut. This is the only carving of this shape found on Palau.

Left: An heirloom stone pounder and bowl from a family in Ngatpang State.

Stone bowls have also been found, often near old stone platforms. Some are just open bowls, while others have stone covers. The one with the cover shown below was found in the jungle of Ngaraard State by hunters. The other open bowl, which is placed on top of stones, is located on a stone pathway in Ngiwal State.

Bottom: This large rectangular stone with holes is located in Ngardmau State. It is called **Orsachel er a Dub,** and the holes on the stone were used as mortars by the men of the village to pound **dub** leaves for fishing. These leaves are poisonous, and kill fish quickly, which was only done when the village had a large feast or funeral.

The large stone coffin shown below is 2.33 meters long, and located next to the road leading from the dock at Ollei, Ngerchelong. The material of the coffin is soft andesite. This is the only stone coffin of that size ever found on Palau, but one small 38 inch long one was also found in the same village.

Hidikata (1956) described the coffin he found in 1939 as being buried in a stone platform, with only a few inches of the bottom of the coffin sticking out above the ground. He opened the lid and found the badly decayed bones of one female person inside, and one small collar bone belonging to a child.

The coffin was later moved to Koror and placed in front of the local Japanese museum. It has recently been moved back to its original location in Ngerchelong. Similar stone coffins have been found in Japan and other countries, usually dating to over a thousand years ago. It makes you wonder why only one such coffin has been found on Palau. The carving of it must have taken several years of work. On a nearby stone platform in Ollei, a small 38 inch long stone coffin has also been found, this one without bones. It was believed to have had moneybeads inside of it.

Above: This now eroded stone statue represents a chicken. It is located at the dock at Ngerchelong State, and believed to have been one of a set of two such statues.

Badrulchau

Also called Bairulchau, this is the largest grouping of monoliths and stone carved faces on Palau, a spectacular site located in Ngerchelong State on the northern tip of Babeldaob island.

The large monoliths were probably foundation stones for a larger structure, since most of them are carved out on top (left) to allow for holding crossbeams of some sort. The stones are of varying sizes and heights, made out of andesite conglomerate.

Other than a variety of mythical legends about a huge unfinished Bai structure, the Palauans have no specific knowledge of the age of these stones or their purpose.

The name means Bai= traditional meeting house, Bad= stone, rulchau= big attraction.

The stone carved faces on these pages are located in Melekeok State. The faces are locally called **Klidm.** The ones with large round eyes are also called **Chesuch** (Owl).

The one shown above is located in someone's backyard, and is half buried. The triangular shape of the stone, and the detailed carving which includes a clearly defined nose, smiling mouth, and earrings is quite different from the rough faces at Badrulchau. It may have been better preserved than some of the other carvings since it has been buried for several decades at least.

In previous centuries, the stones were often moved between villages as spoils of war, so it is difficult to know their original locations, especially for the smaller stones. In the early 1930s, the smaller stones were also moved around and grouped together by the local Modekngei religion followers, to use for ancestor worship rituals and making offerings. Local legends tell of Gods or Spirits who made the stones, and of people turned to stone by the Gods as punishment.

The largest of all stone carved faces on Palau is the one shown on the next page, it is located in front of a house right next to the main road, and is 2.93 meter high. This is one of the more intimidating type faces, with huge bulging eyes and fanged teeth. It is called **Odalechmelech**.

Two other large faces are nearby laying face-down on the beach. Originally there were supposed to have been 9 carved monoliths in this area, forming a rectangular area. Two were moved during the Japanese period in the 1930s, and their present location is unknown.

Left: A smaller eroded carved face is found more inland, not far from the others, and is called **Mengachui.**

This large Owl face is located in Ngeremlengui State, and is a little over one meter high.

Below, left: One of the finest carved faces found, the face has clearly defined eyes, nose and a mouth, and even a hairknot. It is located in Ngeremlengui State, and is still occasionally being used to make offerings to the spirits by some of the villagers.

Below: This face is from Ngardmau State, with barely visible facial features and a phallic shape.

This smiling face is considered one of two sacred stones on Kayangel, the other one is a plain uncarved stone. This carved face was located in front of a house in the village for a long time, but after the typhoon of 1992 it was relocated to stand next to the sacred Hibiscus tree of Kayangel, a tree which the people believe guards them against waves and storms. One family is in charge of the tree and stone, and makes regular offerings. The stone is located in a very dark shady area, and faces the tree, making it very difficult to photograph. Before I was allowed to take a picture, an offering of betelnut was put on top of the stone, and I was told that people had tried to take pictures of the stone before, but they would never turn out good to show the face. (Osborne also photographed this stone when it was still located in the village in 1953).

Located in Ngardmau State, this site with stone pathways and platforms has recently been cleared by the state. The offering table was brought by people who settled here from Ngeremlengui area, and is called an **Olketokel**, used to receive food offerings on ceremonial occasions. Two small holes in the table were used to hang small shells as decoration. Wealthy clans would also use this stone table to line up their Udoud moneybeads, and the clan with the most/ longest bead string would be bringing forth the next chief of Ngerutoi village.

The round stones on top are called **Uleouch el bad**, the name for fried coconut, which was often used as an offering, and to cast spells (**Olai**). When Hidikata saw this stone table and the small round stones around 1929, he did not see the small stone face, which may have been placed with the table at a later date. Osborne photographed the site in 1953, when the carved face was already there.

All carvings shown on these two pages are from different sites in Ngaraard State except the one below from Airai, and the three on this page are unique in that they have more than one face carved on the same stone. The stone on the left has another, larger third face carved on the back of the stone which is not visible here.

Two small stones are known with four faces each. The one shown here below, left, is on the Elab village stone platforms, and difficult to locate in an overgrown area. Aside from the faces, it has a clearly outlined reptile, a crocodile or a lizard, carved on one side.

The other stone below, which is of similar height (19 inches), is found in Airai State, and has four faces with Palauan moneybead bachels outlined under the chins. This stone was photographed by Osborne in 1953 with three of the faces still clearly visible. When I saw the stone in 2000, it was so badly eroded that you could not see any of the faces clearly anymore, and just barely make out the shape of the moneybeads. Osborne's photographs are all that remain of these faces now.

The group of stones in Keklau, next page, may originally have been from different places and placed here in the early 1900s, when a lot of stones were moved around. Hidikata already recorded this group in 1929. It includes two carved faces, one with a hairknot.

Two carved faces are found on Koror and are pictured here. The famous "Mother and Child" stone, left, is found in Ngermid village, and easy to find. The stone pathway leading to the stone is now destroyed, and the stone is very worn and overgrown with lichen and moss. The stone actually has two faces, a small one carved on the forehead of the larger one. The larger face is actually given shape by the carving of a lizard or "child" on the front. Researchers believe the mother and child story developed later, after the original meaning of the stone was long lost, and that the child is actually a reptile shape, as those found on several other stone carvings. This is probably the most intricate and artistic face carving on Palau, with the head of the reptile forming the nose of the large face, while the hollowed out parts form the eyes and mouth.

The other Koror carving, below left, is in Ngerbodel village, on private property on the side of a house on an old stone platform. This is one of the few carvings found which has a head and torso shape, with ear shapes. Half of the face is badly eroded and damaged.

Although other stone carvings have been found in Malaysia and Melanesia, none have an exact or strong resemblance to the Palauan carvings. None of the stones can be dated, and even through association it is not reliable, since pottery and other artifacts in the area could be from later dates. In general, carvings from andesite conglomerate are not well preserved. Most of the large carved faces out of andesite are very weathered and covered with moss. Some others are carved out of harder volcanic breccia and are in better condition, while the faces carved out of coral dripstone are weathered the most.

This large face is located on a stone platform next to the road, before the new Aimeliik Bai area. This is also a complex carving similar to the one on Koror, with a shape of a reptile forming the expression of the large face, and a small face carved on the reptile shape/ nose of the large face.
Below, left: A pair of badly weathered stones found in Ngatpang State. You cannot see the facial features of the large face anymore. The smaller stone has no facial features either, but has a clearly defined neckline and cord around the neck with a Bachel moneybead shape outlined.
Below, right: A phallic shaped stone pillar with lightly carved features found in Airai State near stone platforms.

Aerial view of some of the largest and most spectacular terraces found by Ngchemiangel bay in Aimeliik.

The sculpted hills or terraces found all over Babeldaob are another interesting mystery. Their forms and sizes vary, and according to all the research to date, they were not built for the sole purpose of defense, agriculture, or habitation, none of the evidence is very conclusive. They could also have been large ceremonial areas for the surrounding villages. Most activity seems to have taken place between 1000 to 1400 AD. Palauans have no memory of the purpose of these terraces either. Some of the terraces have stone pathways and platforms associated with them.

Some of the crowns are triangular-shaped, sometimes with a depression in the middle, or sloping up toward the corners like the one shown on the next page from Ngaraard. When I asked the Ngaraard Governor about this interesting shape, he said with a straight face that the chiefs had discussed the purpose of this hill, and had come to the conclusion that it must have been a landing platform for UFOs... so much for oral history.

The earliest mention about the terraces was in 1864 by the trader Andrew Cheyne. He thought they were used for defensive purposes and wrote: “All the hills of the Pelew Islands that are clear of timber are terraced and crowned with a square fort, having a deep and wide ditch around it.” *(Morgan, 1988)*

Ngatpang terraces

Ngaraard terrace top

Ngardmau terraces

Visitors

Although several Spanish, Portuguese, and English ships roamed the Micronesian area in the 1500s already, most seemed to have missed Palau. Probably some of them did spot the islands, or even traded or shipwrecked there over the centuries, but no written records survive. The Philippines, Guam, and the Marianas were known long before Palau. Some believe the Englishman Francis Drake saw the Islands in 1588, but his descriptions of the people and canoes sound like Yap or the central Carolines, not Palau. In 1686 Francisco Lezcano discovered one of the outer islands of Yap and called it Karolina after King Karls II, and the name Caroline Islands stuck for the entire island group ever since.

Canoes occasionally drifted to the Philippines from the Caroline islands, and all the unknown islands east of the Philippines became known as Palaos to the Spanish. A Spanish priest who spent 37 years on the Philippines recorded 8 occasions when canoes drifted in. In December 1696, 29 people drifted into Samar Island on 2 canoes, after 70 days at sea. They came from the outer islands of Yap. After several failed attempts to find the islands, the Spanish sent the ship Santisima Trinidad under the command of Don Francisco de Padilla out to look for the Palaos Islands. They found Sonsorol Island, and some natives came on board, and through the Yapese on board were able to tell them about Panlog or Palao to the North, where they sailed to regularly to get their wood. They tried for several days to find an anchorage, but were unable to because of the strong currents. Two priests and some crew went on land with a small boat, and were never heard off again, since the ship was taken away by the strong currents, and went northeast to look for Panlog. They spotted Angaur and Peleliu, and some canoes came to the ship, but were unable to understand the Yapese on board. The natives tried to steal everything metal they saw, and were given some beads and ceramic items. They started throwing spears at the ship, and were driven off by the crew with gunfire. They describe the Palauans as well-built and totally naked, some with light and others with dark skin, with curly as well as straight hair. They went on to a Yapese island, and then returned to Sonsorol, but were unable to get closer to the island, and the priests and crew did not come out in the small boat. They were forced to return to Manila without them.

Several other unsuccessful attempts were made to reach the islands. In 1712 the Santo Domingo sailed from Guam and tried to capture natives from Mogmog, Palau and Sonsorol, apparently they brought back two men. The French map on page 98 from 1702 shows a vague group of islands "discovered by Drake". The early 1700s map on the inside covers of this book shows the above mentioned 1710 voyage of Don Padilla, Palau is called several names on the map, the New Philippines, Palaos, and Panlog.

By this time there was contact between Palau and the Southwest Islands, and probably also between Yap, Palau, and other islands. The official written history of Palau usually starts in 1783 with the arrival of the Antelope, whose shipwrecked crew spent several months on Ulong Island and interacted with the people of Koror while building a small ship. The Chief of Koror used the Englishmen and their firearms to get rid of his enemies in Melekeok and Peleliu, and Koror became more powerful than it had ever been before. The Captain of the Antelope, Henry Wilson, had his Palau adventures made into a book by George Keate in 1788 after his return to England. This book, which includes the beautiful copper engravings you can see further on in this chapter, became a best seller. It was translated into several languages, and drew attention to these far islands. The key to this success story was a shipwrecked Malaysian who was already on Koror, and who was able to translate between the parties. Without him, things could have turned out quite differently.

In the next century trading ships would pass by occasionally, with Koror always trying to get more weapons and using the foreigners to fight other districts, especially Melekeok. Any visiting traders were strongly discouraged by the Koror Chief from trading with any other districts on Babeldaob, so Koror could stay in charge. Some foreign traders and sailors settled on the islands and married local women, but few lasted very long. Finally in 1883 the British ship Espiegle with the assistance of German researcher Kubary organized a peace treaty between Melekeok and Koror, which was signed by both paramount Chiefs. Foreign diseases had started taking their toll, and the local population was greatly reduced. After 1899 the foreign influence on Palau increased, and under the German, Japanese and American administrations the Palauans lost control over their islands and traditional way of life. The 1788 Wilson book, and early reports and photographs by visiting German researchers like Semper, Kubary, and Kramer, are today still the most accurate and valuable records we have about early Palauans and their traditional way of life.

ISLANDS
F.W.
Hamburgische Wissenschaftl. Stiftung
Südsee-Expedition 1908—1910
II B. 3
Krämer: Palau
4. Teilband
Eilers: West-karolinen
1. Teilband

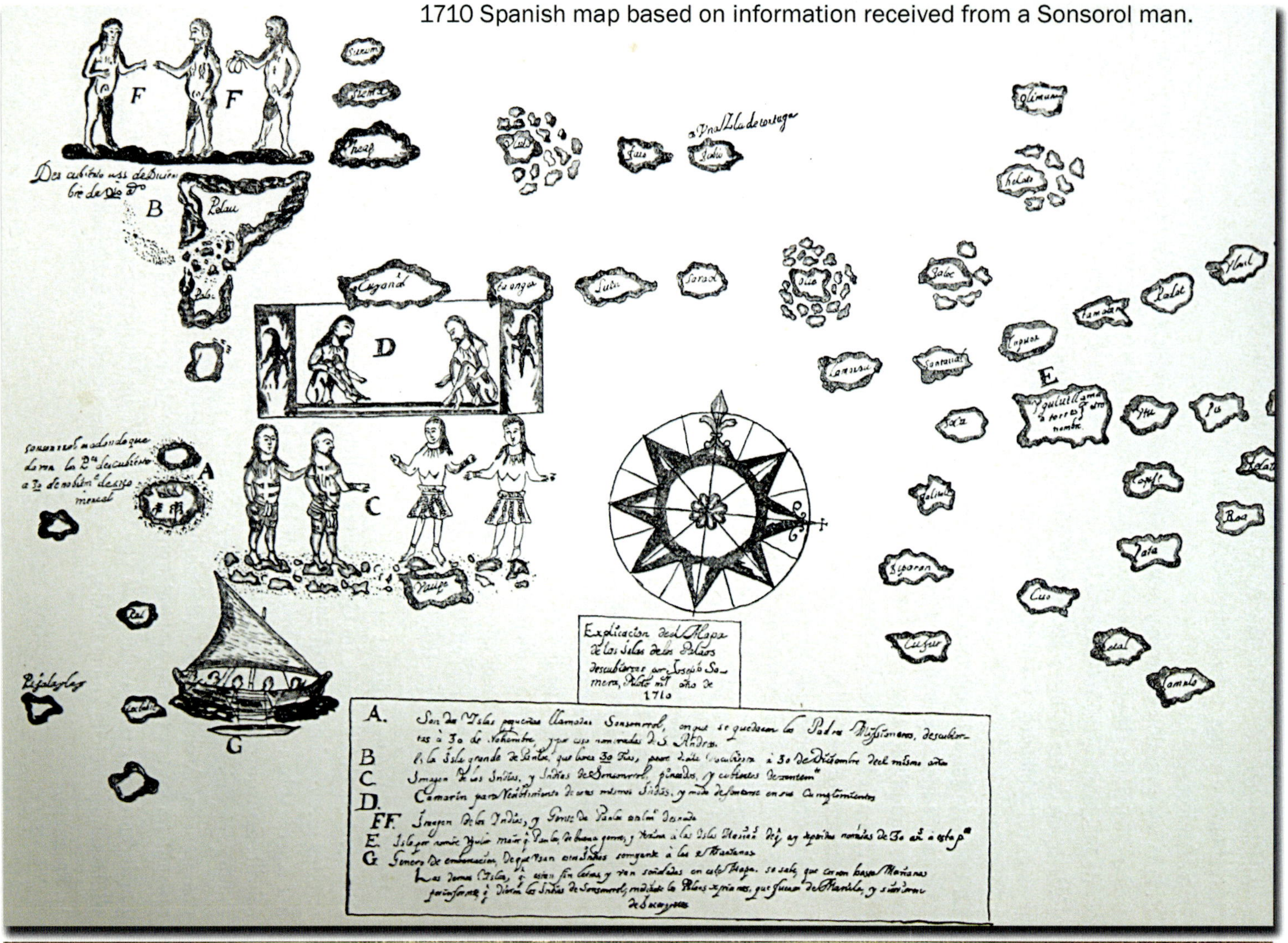

1710 Spanish map based on information received from a Sonsorol man.

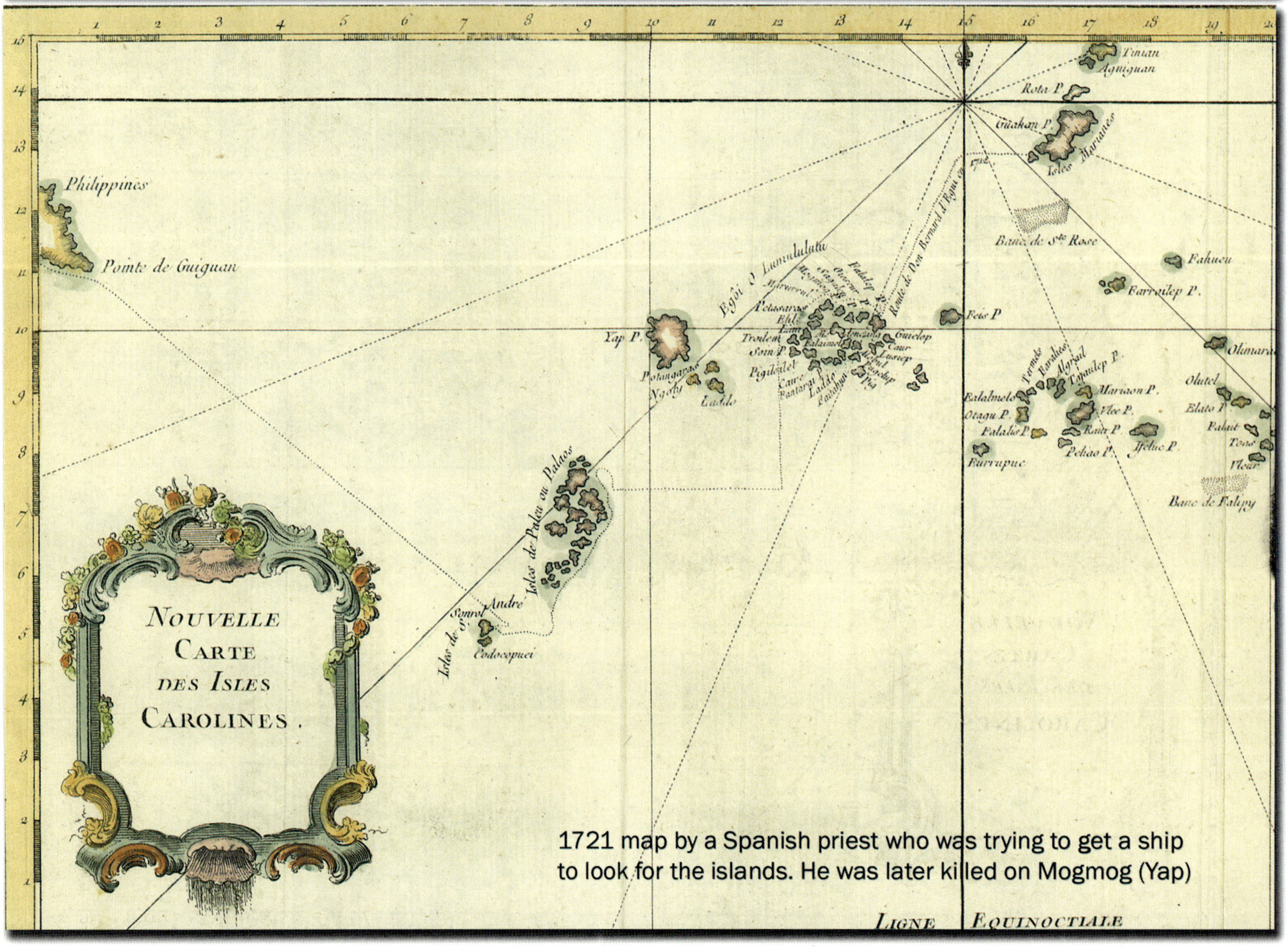

1721 map by a Spanish priest who was trying to get a ship to look for the islands. He was later killed on Mogmog (Yap)

1702 French map

The first regular visitors to Palau that we know about, were the Yapese people, who for some strange reason had decided to start carving their large money discs out of limestone. The limestone on Palau was good quality, so they started coming to Palau regularly. Over the years, they built stone pathways on Palau and traded with the Palauan Chiefs to get permission to quarry, and had considerable interaction with the Palauans. The wearing of loincloths was probably introduced from Yap, since the Palauan men used to go naked.

This interaction started before 1710, and perhaps they already came to Palau for other purposes before the stone money. The Southwest Islands were settled by people from Ulithi, and the village Keklau on Ngaraard was settled by Yapese who drifted to Peleliu. A Yapese man who reportedly brought a lot of moneybeads with him married into the Idid Clan of Koror and made the clan more influencial before the Europeans came. *(Kubary, 1885)*

Left, top: The origin or idea for the Yap stone discs, here seen lined up on Yap, may have come from China. Small discs made from precious stone and jade have been excavated from gravesites dating from 2500 BC, left, and are often associated with beads. They were probably used as ritual objects.

Left: These two stone discs, similar in size and shape to the Yap money, were excavated in Hua-lien county, on Taiwan.

Below, left: Stone money being loaded in Palau on a foreign ship (note the Kotraol canoe in the back) in the late 1800s.

Below, right: Unfinished stone money on an island off Airai.
Below: One of the largest stone money known, off Airai Channel. Another very large piece can be found on Carp Island. Smaller pieces have been found around the Rock Islands, and during construction in Ngaraard, 13 complete pieces were found in the ground recently.
The huge discs are actually some of the later pieces, when the Yapese already had metal tools to work with, and foreign ships from the Irish trader O'Keefe on Yap for transporting the stones. Their value is often less than some of the older smaller pieces, which were carved over a period of years with clamshell tools, and brought to Yap by canoe. The time spent, and the difficulty of the journey home determined the value.

Early undated French print of Yapese canoes, titled: Piroques (Praos) des Isles Carolines.
French print of Yapese canoes in front of Tinian, Mariana Islands, 1827.

A scene from Yap, with a canoe and raft arriving from Palau with two stone money pieces.
This drawing shows two Palauans with a Bai meeting house and spirit house in the background.
Drawings by Franz Hernsheim, 1883.

1783

The ship Antelope, under the command of Captain Henry Wilson, wrecked on a reef off Ulong Island in August, 1783. They stayed there till November 1783, and in exchange for fighting with firearms on his behalf, the Koror Chief helped them built a small vessel out of the wreckage of their ship to go to China, and supplied them with food.

The natives were of a deep copper colour, perfectly naked, having no covering whatsoever; their skins very soft and glossy, owing to the external use of cocoa-nut oil. Each Chief had in his hand a basket of beetle-nut, and a bamboo, finely polished and inlaid at each end, in which they carried their chinam*;* this is coral burned to lime, which they shake out through one end of the bamboo on the leaf of the beetle-nut, before they chew it... It was observed that all their teeth were black (Both sexes black their teeth at the age of 6 or 7 years with a particular gum, and refresh the color from time to time); and the beetle-nut rendered their saliva red, which together with the black teeth gave their mouths a very disgusting appearance.

They were of middling stature, very straight and muscular... and had a particular majestic manner in walking; but their legs, from a little above their ankles to the middle of their thighs, were tatood so thick, as to appear dyed... their hair was a fine black, long, and rolled up behind in a simple manner close to the back of their heads; and appeared both neat and becoming.

The Englishmen were perfectly convinced that the inhabitants had never before seen a white man. They began with stroking the bodies and arms of the English, as if they doubted whether the garment and the man were not of the same substance; they were told by the Malay interpreter that the English were accustomed to be covered... they expressed a further wish to see their bodies... and seemed much astonished at finding hair on their breasts; it being considered by them as a great mark of indelicacy; insomuch that they eradicate it from every part of their body in both sexes.

Our people had 2 dogs, a large Newfoundlander, the other a Spaniel; no sooner had they entered the tent, than the 2 dogs set up a most violent barking, and the natives a noice but little less loud; they ran in and out of the tent, the Malay explaining this to be the effects of joy and surprise; these animals being the first of the kind they had ever seen.

The King *(Chief Ibedul)* was perfectly naked, nor had he any bone on his wrist, or any ornament of distinction. He bore a hatchet on his shoulder, the head of which was made of iron, which much surprised our people, as all other hatchets were of shell. The wives of the Rubaks were rather fairer than the rest of the women, had some ornaments about them, and their faces and breasts were rubbed over with turmeric. Some of the women who have remarkably long hair, let it hang loose down their backs. The women wore only 2 little aprons, or rather thick fringes, one before and one behind, about 10 inches deep and 7 inches wide, which they dyed with different shades of yellow; this they tied round their waists, commonly with a piece of line, the higher rank used a string of some kind of beads, one of this kind a coarse sort of cornelian *(Kau, made from Spondylus shells).* The men had their left ear bored, and wore beads in the ear, and the women had both bored, putting either some leaf through, or an earring of tortoise-shell inlaid. The cartilage between the nostrils was also bored in both sexes, through which they frequently put a little sprig or blossom.

From: An Account of the Pelew Islands, 1788- George Keate

.Wilson Junr. delin^t. I. Landseer, direx

The first Interview between Capt. Wilson & the King of Pelew.

Engraved by R. Reeve

THE END VIEW OF THE PYE A RA MUCKITEE.

The small building to the left of the Pye, under the Tree, is call'd Sum-uck. It is a kind of Religious House, in which the Natives deposit a Wooden Image, call'd by them IDOL ASSACK, and to which they give a portion of Provisions daily

Some of the original Palauan weapons and an inlaid ceremonial sword, brought back to England by Captain Wilson in 1783, now in the British Museum collection.

1783

On August 1783 five of the English embarked in 5 canoes with the King *(Chief Ibedul of Koror)*... in the afternoon, they caught sight of his enemy *(Melekeok)*. The King had with him a fleet of 150 canoes, on board of which were considerably above 1,000 men. Of the enemy's force our people could form no certainty. Before the action the King's brother Raa Kook *(Rechucher)* went in his canoe close to the town, and spoke to the enemy for some time, having one of our men in the boat with him, who did not fire till such time as Raa Kook threw a spear at the enemy. A man was seen to fall directly after firing, and this threw the enemy into great confusion. Such as were on shore ran away, and the greater part of those in the canoes jumped into the water and made for land; a few more muskets were fired, which dispersed the enemy entirely, and in this mark of victory the King being highly pleased with his triumph.

The King had given to Captain Wilson the island Oroolong *(Ulong)*, and in testimony of the King's donation, we hoisted the British pendant, and fired 3 vollies of small arms taking possesion of it for the English.

In September, 10 English went to battle against Artingall *(Melekeok)* again. The King had previously notified to the King of Artingall *(Chief Reklai of Melekeok)* so the latter was not unprepared for the event. A small canoe was sent, containing 4 men, each man having in his hair a white feather stuck upright; the person wearing the feather being regarded in the nature of a herald, either bringing terms or demanding to be heard, hostilities in this interval remained suspended. The enemy upon seeing this also dispatched a canoe, and Raa Kook demanded to know if they would submit to such terms as the King had proposed; the King of Artingall flatly refused. The King then ordered the conch to be sounded; the 10 English were divided in 10 different canoes... each having a musket, cutlass, bayonet and pistol. The English kept up a continual fire, which not only did great execution, but puzzled and bewildered the enemy in the extreme, to comprehend how or why their people dropped... upon finding themselves unequal to so powerful an attack they betook themselves to flight. Six canoes and 9 prisoners were taken, which they accounted a large number. Among the prisoners was a Rubak, who had a bone on his wrist; the people strove all in their power to take from him this mark of dignity, he defended it with singular courage, and lost his life in endeavouring to save it. His head was cut off, and stuck on a bamboo, fixed up before the King's house.

In October, 10 English departed for a third expedition, which appeared to be of great consequence, as all the neighbouring Rubaks were this time to attend the King, and the number of canoes far exceeded those which accompanied the King before. The natives of Artingall behaved with much courage in this engagement. Canoes and 2 houses were destroyed, and also their wharf or causeway, which was much longer and broader than that at Pelew *(Koror)*. Beside doing much other damage to the enemy, they brought away the stone on which the King of Artingall sits when in council.

Again in October, 10 men were to go with the King against Pelelew (*Peleliu*)with a fleet of 300 canoes, our men devoutly wishing this expedition may be the last. A party of Pelew people went to an island belonging to Pelelew, and destroyed the plantations of yams, burnt houses, and cut down cocoa-nut trees. The inhabitants had quitted the island before our people landed. The second evening a council was held, attended by 3 Pelelew Chiefs, and the succeeding morning peace was concluded. Returning to Pelew, the King of Pelelew joined, though in one of his own canoes, carrying 10 women in his train, whether they remained on a visit or as hostages our people were not certain.

From: An Account of the Pelew Islands, 1788- George Keate

1
2
3
4

1
2
3
4

Malton del. after a Sketch on the Spot by H. Wilson Jun.r — W. Taylor, Sc.

The OROOLONG *on the Morning of Departure.*

Published for Capt. Hen.y Wilson 8th April 1803 by G. & W. Nicol Pall Mall.

1783

All our people were employed about the vessel, when their attention was taken off by 2 canoes coming into the harbour to fish for bait; their method of doing it striking all our seamen as uncommon; the boat-men look about for a shoal of that sort of fish which they bait with, not unlike our sprat, these they drive into shallow water, and having covered their canoes and outriggers with matting, they then make a great shouting and noise, splashing the water with their paddles on each side of the shoal, which so frightens the fish, that they spring out of the water, and numbers of them falling upon the matting, are secured in the ends of the canoes.

The King addressing Captain Wilson, said,"You are going, I fear the inhabitants of Artingall will come down in great numbers and molest me, as they have done frequently before; and having lost the aid of your English, I shall be unable to resist them, unless you will leave me a few of your muskets, which you have already taught me to hope you would..." On the boat's return they presented him with 5 muskets, 5 cutlasses, a near barrel of gunpowder, with gun-flints and ball in proportion. In the evening arrived the King's second son, Lee Boo; he was to accompany our people to England. Madan Blanchard, one of the seamen... desired the Captain to tell the King, that he intended to stay behind. The King was very much pleased in the thought of his being to remain at Pelew... and had promised to make him a Rubak, and to give him 2 wives, together with a house and plantations. *(Madan Blanchard was later killed on Peleliu Island in 1790, on Ibedul's order. He had offended many Palauans by not respecting their customs.)*

The little vessel Oroolong was so deeply loaded with their seastores, that a doubt arose whether she would be able to get over the reef, it was therefore agreed to land the 2 six-pounders, and leave the jolly-boat behind. Several canoes preceeded the pinnace, to mark the safest track for the vessel... the King accompanied the English in their vessel almost to the reef, before he made a signal for his canoe to come alongside.

An accident, wholly unexpected hath given us at last a transient sight of these people; nor is it likely that they will again be sought, as they posses nothing but good sense and virtue, and live in a country which supplies no materials that may tempt mankind to disturb their tranquility.

After leaving Palau, the English made it to Macao in their little boat, and proceeded to Canton, where they embarked for England onboard the ship Morse with the Palauan boy Lee Boo.

From: An Account of the Pelew Islands, 1788- George Keate

1788 drawing of the camp on Ulong.
An aerial view of the area on Ulong Island where the English had their camp and built their small boat.

PRINCE LEE BOO,
Second fils d'Abba Thulle.

1784

He *(Chief Ibedul)* had resolved to entrust his second son, whose name was Lee Boo, to Captain Wilson's care, that he might have the advantage of improving himself by accompanying the English, and of learning many things that might at his return greatly benefit his own country. The young man was remarkably clean in his person, washing himself several times a day... being impressed with the idea of the indelicasy of having no clothing, he never after appeared without his trowsers...

Lee Boo, upon seeing the large Portuguese ships at anchor in Macao, appeared to be greatly astonished... upon seeing the first house our young traveler had ever seen, what struck his imagination at first, was the upright walls and flat ceilings of the rooms, he seemed as if puzzling himself to comprehend how they could be formed. The vessels of glass appeared to be the objects which riveted most his attention... a large mirror at the upper end of the hall reflected almost his whole person. Here Lee Boo stood in perfect amazement at seeing himself; he laughed, he drew back, and returned to look again, quite absorbed in wonder, he made an effort to look behind it...

Our people bought him some little trinkets, which they thought would from their novelty please him. Amongst them was a string of large glass beads, the first sight of which almost threw him into an ecstasy; he hugged them with a transport that could not be exceeded by the possessor of a string of pearls of equal magnitude.

Whilst at Canton, several gentlemen who had been at other places, where the throwing of the spear is practiced, and who themselves were in some degree skilled at the art, having expressed a wish to see Lee Boo perform this exercise; Lee Boo at first did not point his spear to any particular object, but only shook and poised it, this they were also able to do; proposing him to aim at some point, they fixed this to be a gauze cage which hung up in the hall, and which had a bird painted in the middle; Lee Boo took up his spear with great apparent indifference, and leveling at the little bird, struck it through the head, astonishing all his competitors, who at the great distance from where they flung, with much difficulty even hit the cage.

He was extremely desirous of knowing the names and country of every ship he met at sea, and would repeat what he was told over and over again till he had fixed it well in his memory; and he made a knot in his line each time, but these knots, now having greatly multiplied, he was obliged to repeat them over every day... and often to recur to others, when he had forgot what a particular knot referred to. As we drew near the British Channel, the number of vessels that he observed increased so much, he was obliged to give up the keeping of his journal... *(They arrived in England July, 1784)*

He was requested to give us a Pelew song, and obligingly began one as soon as asked; the tones however, were so harsh and discordant, and his breast seemed to labour with so much exertion, that his whole countenance was changed by it, and everyone's ears stunned with the horrid notes. From this sample of Pelew singing, it is not to be wondered, that a chorus of such performers had the effect of making our countrymen at Pelew fly to their arms!

On the 16th of December, Lee Boo felt himself much indisposed, and in a day or two after eruption appeared all over him. Captain Wilson called to inform me of his uneasiness... apprehending that it might be the small-pox.
(Lee Boo died and was interred in Rotherhithe church yard in England, where his tomb can still be seen today.)

From: An Account of the Pelew Islands, 1788- George Keate

To the Memory of

Prince Lee Boo
A Native of the Pelew or Palos Islands
and Son to Abba Thulle, Rupack or King
of the Island Coorooraa;
who departed this Life on the 27th
December, 1784
Aged 20 years;

This Stone is inscribed by the
Honourable United East India Company
as a Testimony of Esteem
for the humane and kind treatment
afforded by his Father to the Crew of their
Ship the Antelope, Captain Wilson, who
was wrecked off that Island
in the Night of the 9th of August 1783

Stop, Reader, stop!
-Let nature claim a Tear-
A Prince of mine, Lee Boo, lies bury'd here.

1791

The East India Company in 1790 sent two ships under the command of Captain John McCluer from Bombay to Palau, to inform Ibedul of his son's death, and to stock the island with cattle, seeds and plants from Indonesia. Two of the officers had been part of the Antelope crew on Palau. They arrived in Palau in 1791, and one of the ships took a 4 month trip to Macao with 3 Palauans. Another two year trip was taken to New Guinea and Indonesia with 3 Palauans on board, but 2 got sick and died. In the mean time McCluer raised cattle and planted on Palau, and assisted Ibedul with his local warfare.

When they came in sight of the landing at Cooroora *(Koror)*, the conch shell was sounded to announce their approach; and the people in the canoe abreast of the King's began a song, the first line of each verse being given out by the King; the verse ended with a general chorus, accompanied by a flourish of paddles from all canoes at the same instant, which had a most grand and pleasing effect.

When an inferior Rubak displeases the King, he takes from him his bone *(Dugong bracelet),* and either keeps it or bestows it upon some other person. The practice of tattooing and also the blackening of the teeth, is universal with both sexes; the females of high rank have it done at the age of 6 or 7 years. The hands undergo the operation first, then the arms as high as the shoulders, after which the feet, upwards to the hips. Among the lower class of women, the time depends on their abilities to pay the female artist; but as it would be disgraceful to marry before this operation takes place, it is rarely delayed to years of maturity. The men puncture at any time, and as much as they please; in general however, the hands and one thigh are done, with a line up the arms representing birds, fish, etc.

Their marriages are without any public form or ceremony, excepting that of obtaining the consent of the parents or chief Rubak... a Rubak or any other person may have as many wives as he can maintain, but each wife must have a separate house; this custom therefore gives the richer Rubaks more wives than the common people.

The Bais, or public buildings, are the most astonishing fabrics that we ever beheld; considering the tools and workmen that constructed them. Since the loss of the Antelope, they have built a new Bai of near 60 feet in length. The floor is a perfect level, and many of the planks are from 3 to 4 feet broad, joined so close that a pin could not pass between them; the breath of this house is nearly one-third of the length, the windows resembling the port-holes of a ship, 6 or 8 exactly opposite to each other; the beams are laid about 7 feet high above the floor, curiously carved, and so nice fitted to the supporters that they appear as one piece of timber; the roof is high, having a great slope; the thatching is a most ingenious performance; being done with the cocoa-nut leaf; the inside of the house was most curiously worked and ornamented with various figures and flowers, the ends had much the appearance of Gentoo temples. Every Rubak or chief has a square piece of pavement before his house, and a small detached building, like a pigeon-house, which was taken for a place of worship.

There being plenty of water at Amallikala *(Malakal)*, in the harbour of which the ships are now secured, the King preferred residing here to his capital Cooroora... This morning the King had given the English this island, saying it should be Englishmen land, on Captain McCluer's haven spoken in praise of it, its being so plentyfully supplied with springs and running water; the soil being good and fit for cultivation. The English colours were now hoisted upon the point of the island; the foundation stone of a fort was also laid, to be called Fort Abercrombie. This island is not above 4 or 5 miles in circumference...and good anchoring ground all round it.

The ships were about to leave in 1793, when McCluer resigned and decided to stay on Palau. After 15 months he left Palau with his Palauan wife and son, 6 Palauan women and several men to Indonesia, from where he sent his family and the 6 women on to Bombay. He himself and the other Palauans sailed off from Bengal, and were never heard from again. In 1797 the three remaining Palauan women were living on Bombay with a sailor, and requested passage home, after which Captain Wilson brought them to Macao, from where they hired a small boat and finally returned to Palau.

From: The Journals of the Panther and the Endeavour, 1791-1793.
Supplement to the Account of the Pelew Islands, 1803- John Pearce Hockin

Habitans des Iles Pelew

Coconut moneybead holder with shell inlay and coneshell cover. British Museum Collection.

1791

We got into a good anchoring place the 11th of June, when our vessel was crowded with natives of both sexes, and the water was covered for 40 or 50 yards round with canoes... they came on board with great eagerness, and with as little fear or suspicion as a child would enter a parent's house...

It ought to be stated that the King always gave notice to an enemy at least 3 days before going to war with him... as Captain Wilson had done before us, we joined the King against the inhabitants of Artingall... some thousands of men were embarked. I was assigned to the command of the launch, a large boat with a crew of Europeans. We had a six-pound brass cannon, swivels, a chest of ammunition, and each man a musket... the canoes formed 3 lines, front, center, rear. The launch with English colours flying, was in the center...

With our spy-glass, we saw that the Artingall beach was covered with natives for a quarter of a mile near to the town, who had arms in their hands... we observed people going to a stone pier and enter a canoe, which was paddled directly to our boat, at the astonishing rate of 8 or 9 miles an hour... they drew up at about 4 yards distance from us, and at the clapping of hands by the sternsman, they all at once backed water with their paddles, and stopped as suddenly as if they had struck a rock... we saw the Chief who was with them, and who sat distinguished from all the rest upon a seat in the center. Their conduct upon this occasion attracted my attention and excited my admiration... no signs of fear or doubt were betrayed by them, notwithstanding our expedition and various arms must have appeared formidable to them, unaccustomed as they were to meet a foe thus equipped.

The King said to the Chief, "Are you ready to fight?" He answered,"We are not, but we will sooner fight than have any laws imposed upon us which we think unjust and disgraceful." The terms which the King proposed (and which were later accepted) were that the people of Artingall should carry him from his canoe on a kind of litter to their place of state, and set him on the throne: that the 2 highest chiefs who had been named King of the 2 islands in rebellion should bring him several valuable jewels *(moneybeads)* which they held at that time and which had descended from his ancestors; that they should acknowledge him to be their lawful sovereign and promise never to revolt on pain of death; that the lower chiefs should prostrate themselves before him with their faces to the ground and make the same promise; and that they should exchange 60 women as hostages to secure the observance of peace...

This was not the first time that our people had gone out with the King against the people of Artingall. The crew of the Endeavor, now with us, had been engaged in the same way. The officers told me that men from Artingall had fought terribly, and that no people were more brave. The inhabitants of Pelew had taken several of them prisoners, and had put them to death in a manner not agreeing with their general character for humanity...it is certainly inhuman to cut off the leg of a prisoner and beat his own face with it while he is yet living... The maxim handed down from their fore-fathers was, that more is to be feared from one prisoner than from five open enemies...

From: Narative of Voyages and Travels, 1817- Amasa Delano

SEP(150)TEN
160
170
180
COREA
Pinggau
I.Ochi
Achita
Note
Nambu
IEDO
Meaco
INS. NI PHON
I. Mataixima
I. Toy.
I. Yynoxima.
INS. IAPONICÆ.
I.XICOCO
Satzuma
I.XIMO
I.Tanaxima
Vencheu
I.de Fuego
I.Lequeio Grande
I.dos Reyes Magos
I.Formosa
I.dos Herman-nos
I.Mal abrigo
I.Amsterdam
I.Desierta
I.dos Columnas
CANCRI
TROPICVS
I.Vrac.
I.Agrigan
I.Pagon
S.Xaverij
S.Ignatij
I.Amalagan
I.Guguan
I.Sarigan
I.Anataian
Conception
S.Philippi
S.Caroli
S.Ioachim
INS.MARIANÆ,
ol. Latronum
I.Saypan
S.Ioseph
I.Tinian
I.Zarpana
I.Guaha
S.Angeli
S.Annæ
S.Ioan
I.Pintados de Paßaros
INS.PHILIPPI-NÆ.
I.Tandaia
Achan
OCEANVS
ORIENTALIS.
I.Carolina
I.S.Barnabæ
INS.de PAIS.
I.Lamures
Mindanao
S.Iuan
I.Palmeiras
I.Arecifes
I.Matelotes
P.º Villan
I.de los Saldadores
I.S.Iuan
LAGVS
ZARI
ARCHI-
PE
LA
I.Morotay
I.de los Martyres
I.des Aues
LINEA
ÆQVINOCTIALIS
I.CELEBES
Insulæ Moluccæ
GILO
Terra dos
PAPOVS
Hoch Land
Legh Land
NOVA
GVINEA.
I.Ceram
I.Banda
I.Aru
I.Timor
I.Timor
Leucæ Germanicæ.
Leucæ Gallicæ.
Leucæ Hispanicæ.
MERI(150)DIES
160
170

From: Historia de los Trages que Todas las Naciones del Mundo 1804- Madrid

1793

We arrived back at the Pelew Islands the 20th of January, 1793. We were pained when we found the confusion and the sufferings which the misconduct of the successor of Abba Thulle *(Chief Ibedul)* had brought upon his people. The people... were now in a state of oppression, mutual distrust and opposition... The firearms and ammunition which we had left with the Pelews, had done them incalculable injury. They were divided into parties, and were frequently at war with each other. Unhappily for the Pelew Islanders, they have lost much of their early simplicity and goodness, and have not yet gained the intelligence and virtue of a civilized people. It is a wise provision of nature that savages should be limited to few and simple weapons of warfare till they have acquired the habits, and have entered into the pursuits of civilized society. This system of things ought not to be violated by us...

A circumstance on our arrival will show the wonderful activity of the natives of Pelew in managing their canoes... as we approached the town, a canoe having a sail and outrigger with 10 or 12 men, came off according to custom to meet us. We were going at a rate of 8 miles an hour. We threw out a rope for them, which they caught while in the center of the canoe, not knowing the effect of it upon themselves. The canoe was immediately turned over, and they let go the rope. They were excellent swimmers. They got their canoe upright at once, bailed out the water, arranged the sail, and with the wind and paddles overtook us at a distance of a mile...

Their canoes always have one outrigger. Some of those for war are 30 and 40 feet long... they can go equally well with either end first, and always keep the outrigger to the windward in sailing... The canoes are beautifully ornamented with inlaid work of mother of pearl, or of any shell which is white. The outside is always painted red, and the white figures are very conspicuous by the contrast, imitating birds, fishes, flying foxes and men. The paddles are often adorned in a similar manner. One man of the crew pronounces a kind of chant, and instantly they all flourish their paddles over their heads with a perfectly uniform motion, and with the greatest dexterity.

The children of the Pelews are all taught to swim very early, the girls as well as the boys. I have seen the mothers throw them into the water 20 times in succession, when they were not more than 2 or 3 years old; and catch them up again after they had paddled awhile with their hands and feet, and were about sinking. They are remarkably sharp sighted... the mothers are in the habit of squeezing green lime juice into the eyes of their children when they are a week old, and afterwards. I have seen this done often, and even to adults.

It was considered as an honour for any woman, married or unmarried, to be in a state of pregnancy; and if she were unmarried, it recommended her to a husband. Marriage was esteemed among these natives as a relation of great importance and responsibility.

I have seen my brother Samual during the fall of 1816, who tells me that... the islands seem to be in a better condition... some Europeans have settled there, taking wives according to custom from among the natives, and raising a population of mulattoes.

From: Narrative of Voyages and Travels, 1817- Amasa Delano

Hombre de las Yslas de Palaos

1832

The American whaler Mentor, under the command of Captain Edward C.Barnard, hit a reef off Kayangel in May, 1832, after a 3-day storm. Half of the 22 crew died the first night while trying to get away from the wreck with a small boat.

Before sunrise we discovered a canoe within a short distance of us, containing inhabitants of the neighboring island *(Kayangel)* ... their appearance excited my astonishment. They were entirely naked. Each one was armed with a spear and tomahawk; some had battle-axes. They were fantastically tattoed on different parts of their bodies. Their hair, naturally coarse and black, was very long and hung loose over their shoulders. Their teeth were entirely black... Just at this time there came in sight a number of other canoes, perhaps 30, filled with natives, who seemed no less intent upon plunder than those with whom we had already formed a disagreeable aquaintance. After a while they offered to render us some assistance by towing our boat... no sooner was this arrangement completed than a Chief and one other native took their station with us... and seizing our powder canister, the Chief jumped overboard and swam to his canoe. His companion seized a bundle of clothing and was making off with it... Mr. Nute took a knife and deliberately cut the line, our intention... to escape from them... this we succeeded in accomplishing.

After they left us, rowing until about sunset, we discovered land ahead *(Babeldaob)*. We continued to row on, and at about 4 o'clock the next afternoon we succeeded in landing on a small island distant from the main land about half a mile *(Ngerechur Island, uninhabited)*. A canoe with 2 men in a state of nakedness was seen... we presented to each a jackknife ... they seemed highly gratified, their conduct was inoffensive... we followed them towards a sort of harbor at no great distance *(Ngerchelong)*. A large war canoe made directly for us; and upon coming alongside, the head Chief sprung into our boat, seized the Captain by the shoulder, and struck him several times with a war-club... he began to strip us of our clothing... we were soon in their miserable harbor. We were conducted to a platform, on which were seated those who had power to dispose of us as they pleased. This platform was situated between 2 long buildings called pyes *(Bais)*, a messenger was dispatched to a neighboring town to consult their prophetess... and she had directed them to send us to her.

We had not long been with them before we became aquainted with the fact that there was another tribe *(Koror)* and that they were not on friendly terms with each other. Intelligence had in some way been communicated to those who lived remote from where fortune had thrown us... they sent to us a message of their willingness to assist in constructing a boat... the persons commissioned to make this proposal were 2 Englishmen, who had been on the island for several years. An offer of that kind, coming as it did from their enemies, and being in itself calculated to offend the pride of those into whose hands we had fallen, greatly excited their feelings of animosity... previously to this, some steps had already been taken towards construction a sort of boat or vessel... Our actual wants were supplied; and the natives soon evinced a disposition to consider us friends, and treat us as such. We were indebted to them for our maintenance while among them, and for the assistance they rendered us in fitting up our craft... situated as we were, we did not feel ourselves at liberty to expostulate against the obvious unreasonableness of their demands... we solemnly assured them we would send them 200 muskets, 10 casks of powder, balls and flints... articles of ornaments... Provisions were furnished us by the natives; but we greatly needed a compass. Captain Wilson who had been shipwrecked there many years before, left his compass with one of the Chiefs, whom we finally succeeded in inducing to part with it. On the 23rd of November, we set sail... it was agreed that 3 of our number should remain as hostages, and 3 of the natives should accompany us.

The 8 crew and 3 Palauans drifted to Tobi Island, where they were helt prisoner with little food. Four crew and 2 Palauans died there. The Captain and 3 crew escaped on passing ships in the next 2 years. The surviving Palauan on Tobi, and the 2 hostages on Ngerchelong were finally rescued in 1836 by a US Navy ship. (The 3rd hostage had also escaped on a ship)

From: A Narrative of the Shipwreck... 1836- Horace Holden

1740 copper engraving of a strange looking shark and a remora.

Left and Next page:
Drawings from a German children's book, based on the adventures of Karl Semper on Palau.

1860s

Earlier I had learned that for some years district wars had been flaring up as a result of the arrival of the traders Woodin and Cheyne... At first in 1860, both of them stayed at the Corore *(Koror)* harbor, and for a time they had done business together. Corore now feared they would lose some of their power over the other districts if Woodin would start trading directly with Uibukit *(Ngabuked, Ngaraard)*. Corore had enjoyed this power since Wilson's time, as a result of the assistance he had given them to overpower the other districts.

King Ibadul of Corore went on board Woodin's boat, and explained to him that they had come to punish the inhabitants of Uibukit for having assumed trading rights which did not belong to them. Woodin stood his ground...fortunately, the battle did not become a slaughter...a canoe commanded by a courageous young chief from Uibukit came so close to the canoes from Corore, that one shot fired by him put a hole in a Corore canoe so that it sank at once. the whole southern fleet turned to flee from the pursuing enemy. With this one victory Uibukit had won a position such as it never had before. Several minor districts put themselves under its protection at once, so that Palau was now divided into two groups.

In January 1862, in the absence of Woodin and his ship, the village of Uibukit had again been exposed to an attack by Corore. this time, Cheyne's allies (Corore) had had the active assistance of an English Man-O-War. A number of canoes were stolen, and they had burned several houses, but Uibukit's standing had not suffered much in spite of this blow, since Woodin now returned with me.
(The Palauan people considered Semper as powerful a foreigner as Cheyne, and all the locals including Ibedul expected him to call up a warship also, to take revenge on Koror).
Ibadul now came and offered his former enemies peace and friendship. The stolen canoes and items were not returned, but a large piece of native money of the most valuable Brack *(Berrak=yellow)* type compensated for the loss completely.

From: Die Palau-Inseln im Stillen Ocean, 1873- Karl Semper

In 1862 Semper found a contract and petition Cheyne had drawn up, signed by Chief Ibedul and the Koror Chiefs, to give Cheyne exclusive trading rights on Koror, and also claiming Koror's authority was being threatened if the other districts received more firearms. This was brought to the British Secretary of State in Hong Kong. There was also a draft "Constitution of the Pellews". The British government never acted on this, and Cheyne was murdered on Ibedul's order in 1866, after he had started trading with other districts and bought land on Babeldaob himself, and had given firearms to Melekeok. The British investigation into his death took place 14 months later, and they shot the Ibedul of Koror as punishment for the crime. Local stories say a different Koror Chief impersonated the Ibedul, and that the wrong man was shot. Another trader who worked with Cheyne on Palau in those days was Alfred Tetens, he lived on Palau from 1862-1868, and left to Europe after he was shot in the leg while helping Koror in a war against Melekeok.

Abenteuer
eines Naturforschers
in der Südsee

1876 photograph of the Chief Ibedul Chokebai Ngirailengelekei and his family. He became Ibedul in 1871.

1874

The German trader Edward Hernsheim arrived in his own ship, bought some land on Malakal, and started a seacucumber trading post on Palau and later on other Micronesian islands. While in Hong Kong in 1875, the Spanish consul demanded taxes from him for trading on Palau, claiming the islands as Spanish possessions, but the British and German trading on the islands continued.

On the beach of Malakan *(Malakal)* we were greeted by the Dutch trader, who runs a trading post for a German company here. Formerly a sailor, he has been on this island for many years, and often does not see a white man for six months at a time. He has two huts, one for trading, and in one of them he stays with his wife, who is a daughter of the highest Chief. As proof of their special friendship, her father sold her to the Dutchman some years ago for 2 muskets and a metal ax. She speaks a little English, wears a colorful wrap, and only puts her cigarettes away to put a refreshing betelnut in her mouth. The Dutchman has to take care of his kitchen himself, but his wife does help him with some light tasks, such as fetching water and opening coconuts. He is a funny one, this Dutchman, he at all times has a dog, a goat, and a cutely noisy little pig following him around, and a tame parrot that rarely leaves his shoulder. He also has a large black cat that sleeps next to his gin bottle on the table.

Shortly after our arrival came the Aba-Thule *(Chief Ibedul of Koror)*, a big man with dark brown skin, shiny with oil. We entered his house, and surrounded by his women and chiefs, Aba-Thule sits on the bamboo floor. They were all busy chewing betelnut. A hand movement from the King shows us a place near him to sit... Our meal was served on artistically shaped tortoise shell plates and leaves, and it gave me time to look around... All sorts of things are hanging on the blackened walls and standing around in a strange colorful mixture, you get the feeling of being at a flea market. An old rusted rifle hangs next to a shiny new one, gifts from ship captains, and both unusable. The first was broken, and the second was given with only 6 bullets out of safety concerns, and he wasted those 6 the first day. A broken mirror frame holds a page of the "Illustrated London News"as a painting, empty bottles, metal pots, an old ship cannon, a painted chest, binoculars, and an empty clock cover complete the colorful collection... Aba-Thule is finally satisfied, and ate in proportion to his enormous body size, and now it's back to the betelnut...

Only in the evening when Aba-Thule brought us to the Bai did we see young girls, coming from the fields in long rows. They were carrying baskets with taro on their head, bundles of sugarcane, coconuts and bananas. Their big eyes laughed and looked wonderingly at the strangers, their hair decorated with fresh flowers and held up in a knot, their cute yellow grass skirts just covering the hips. We noticed the uniformity of their beautiful round shapes, covered in turmeric and oil, the glow of the setting sun giving them a bronzed look...
Soon, young men arrived with large baskets of speared fish, and all sorts of other strangely shaped inhabitants of the salt water... After they emptied the tortoise shell plates and leaves, they went back to cigarettes and betelnut, and pairs started forming, who, according to the local custom, whenever they feel like it are allowed to spend a night of free love in the Bai...We did not notice the time, and it was too late to return to the ship, so we gladly accepted the friendly invitation of Aba-Thule to spend the night. Unfortunately the ship was ready to sail the next morning. Aba-Thule sent us 6 large canoes to guide us through the reef. For a few more minutes did we hear their cheerful singing, and then the deep silence came, the one only those who have sailed through tropical seas for long periods know and learn to love.

Text and drawing next page from: Sudsee-erinnerungen, 1883- Franz Hernsheim

1880s

Johann Stanilaus Kubary came to Palau in 1871 to collect for the Hamburg Godeffroy Museum, and again in 1883 to collect for the Leiden Museum. He married a half Pohnpeian girl, lived on Koror and in Melekeok for a while, and was even made a number 4 chief of Melekeok. He was able to collect several pieces of Palauan moneybeads during his stay (see next page). His influence grew during a bad flue epidemic, when he cured several Chiefs. A ship from Pohnpei to Germany with 100 crates of his collections sank, and his plantation on Pohnpei had problems. Kubary started drinking, and after living in New Guinea for several years killed himself in 1896. He made some of the first photographs and in-depth observations of Micronesians. His photo of a two-story Bai in Airai is the only evidence left of such a structure in Palau.

Melekeok had gained power through its ruling clan Udes, and Artingal became the largest and most powerful district in the islands. Other districts came under its rule, until Koror, with the help of English firearms, brought the other districts to its side and weakened the power of Melekeok. Kubary in 1883 says:

With the help of the white men, Koror has suppressed and partially annihilated the peoples of the northern island, although their civilization was much more highly developed than that of the natives of Koror. Traces of this remain visible today, and the proof is the fact that the lazy and indolent Koror is supported solely by our ships and never had an industry. The money, earthenware, canoes, nets, oil, syrup, pigs, etc, all of it came and still comes from the north.

In 1881 the British warship Lily gives an unreasonable fine to Melekeok, because people from Ngchesar had plundered one of the trader O'Keefe's ships and treated the crew badly. The fine was 2000 dollars in copra, 1000 dollars in turtleshell, 1000 dollars in oysters, and 1000 dollars in seacucumbers. Half the payment was delivered after 6 months but refused, and the next year the Lily was back at Melekeok with one of Ibeduls helpers, James Gibbon, who now demanded the fine to be paid in moneybeads instead. Reklai came up with one one piece, while they expected 30. Another British warship came up from Koror, and the 3 Bais of Melekeok and Reklai's house were burned, and so were Bais and a priest house in Ngchesar. In 1883 the British ship Espiegle under Admiral Bridge came to check out the situation, and lessened the fine. He also drafted a peace agreement between Melekeok and Koror, and forced both Chiefs to sign it. (see next page) Kubary's comment:

He (*Admiral Bridge*) deserves credit for being the first to try and set right the terrible mistakes of those who came before him, through a noble deed done on his own initiative.

Around this time, the Irish trader O'Keefe started using his ships to transport Yapese stone money from Palau to Yap in return for copra. One of his agents on Palau was a man called James Sims from Singapore, who married a Bai girl from Angaur, and by bringing in firearms he helped Angaur become independent of Koror. In return they made him a Chief on Angaur, where he lived for many years.

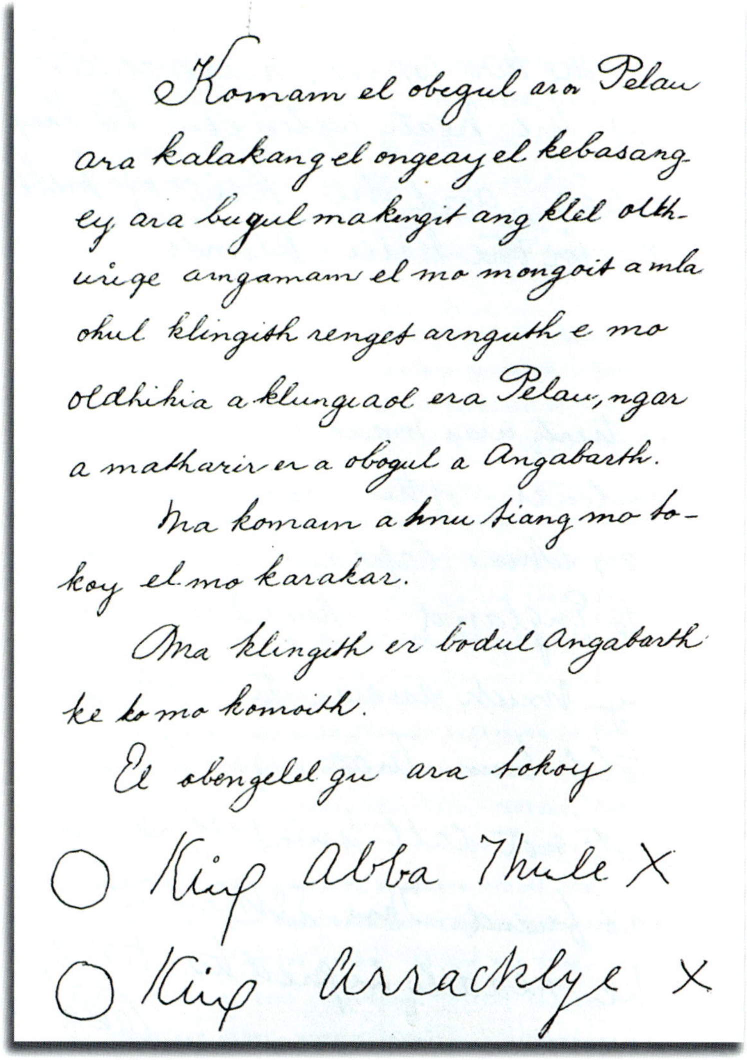

Komam el obegul ara Pelau
ara kalakangel ongeay el kebasang-
ey ara bugul makingit ang klel olth-
uiige amgamam el mo mongoit a mla
ohul klingith renges arnguth e mo
oldhikia a klungeaol era Pelau, ngar
a matharir er a obogul a Angabarth.
Na komain a hnu tiang mo to-
koy el mo karakar.
Ma klingith er bodul Angabarth
ke ko mo komoith.
El obengelel gu ara tokoy
O King Abba Thule X
O King Arracklye X

1883 peace treaty between Koror and Melekeok

Kubary's drawing of a Bai in Melekeok, where he lived,

Palauan moneybeads collected by Kubary.

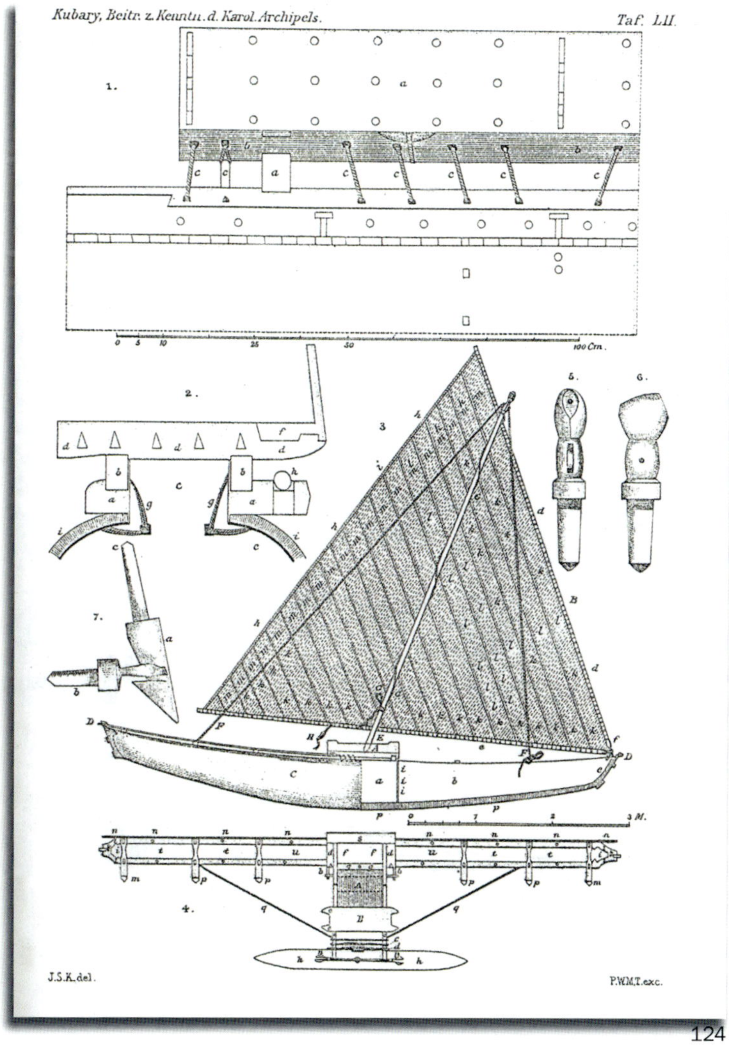

and some of his blueprint details of Palauan canoes.

Augustin Kramer

1908-1910

The German Sudsee Expedition that took place between 1908-1910 had researchers visit different islands to study the local cultures, and the expedition covered all the Micronesian islands. The detailed descriptions, drawings and photographs made during this two year time-span, when most of the islands still lived in their traditional way, but were on the verge of change, are still very important today. Augustin Kramer and his wife Elisabeth were in Palau twice during these two years, for a total of nine months, in between they visited and studied the outer islands of Chuuk. Five book volumes were published on Palau alone, with an incredible wealth of accurate information on the people, the culture, and with important details such as blueprints of the Bai structures and canoes, as well as Palauan names, songs and legends. A lot of cultural information would have been lost if it wasn't for these books, because of the enormous changes that the Japanese, WWII, the Americans and missionaries brought to the few Palauans left.

In the year 1905 I decided to visit Kubary's favorite area, Palau. The excellent work he collected and published here and there screamed for completion. I first planned a short visit, and set myself the task of recording the strange decorations that show on the Bais, since Semper and Kubary pretty much ignored them, but soon found out a lot more work needed to be done... How good, that she (Elisabeth Kramer) captured all important images and decorations in color, because now a typhoon in 1912 destroyed almost all Bais, and a lot will probably not be rebuild. My camera and her pencil made sure they were not lost forever. The details and explanation of the images was done for me by my very smart helper William Gibbon, whose West-Indian father is often mentioned in the early traders' stories. I could speak English with him, and help him improve his German, which was a pleasure.

William Gibbon kept in touch with Kramer after 1910, and sent updates and artifacts from Palau to him. the fifth and last book volume was published in 1929.

I took on this assignment on the condition that my wife could join me. Not only was drawing and painting her field, but with the research of the women on Palau I owe her so much I should actually put some of the chapters under her name... That is why I dedicate this book also to her. What we went through on this expedition as far as danger and suffering, does not show in these books. How many hundreds of boat rides we have taken together in wild seas, over shallow reefs and in strong currents, in rain and unrelenting sun... We wandered over hot stones, through swamps, over cliffs and through the jungle, and at night, our bed was a small hut, but the hard wooden bottom was a welcome bed nonetheless, with a mosquito net being the only separation between us and the outside world. All this does not show in these books. Also, it does not show how much we enjoyed the beauty and the excitement...

From: Ergebnisse der Sudsee-expedition 1908-1910
Augustin Kramer: Palau, volume 1, 1917.

View of the stone pathways of Ulimang village with a Spirit house (left) and a Bai, in Ngaraard around 1907.
A Bai, boat house and canoe on Koror.

Watercolor by Elisabeth Kramer- Bannow 1907
Dirangeaol from Koror and Kelebil from Melekeok

Watercolor by Elisabeth Kramer-Bannow, 1907. The Ngermid Bai on Koror is where most foreign visitors stayed, from Captain Wilson to Kubary.

The town of Koror in the 1930s during the Japanese occupation.

During the German period (1899-1914) the Japanese were already trading copra and doing commercial fishing on Palau, in spite of efforts by the German government to stop them. When Japan declared war on Germany in 1914, Japanese naval forces arrived on Koror, and Palau became one of the naval districts under command of a Japanese Admiral stationed on Chuuk.
In 1919 a mandate for the islands was awarded to Japan, and by 1922 the civilian South Seas Government was established, under a Governor. Koror became the administrative headquarters for Japan in Micronesia, and in the 1930s, around 30,000 foreigners lived on Koror. Paved streets, electricity, shops and a water and sewer system changed the look of Koror.

Public schools, required by the mandate, were started only through third grade, and medical facilities were improved, although separate facilities were used for Palauans and Japanese. Research facilities were set up to study the feasibility of agriculture, fishing and mining on Palau. Bauxite and phosphate mining, copra, and local crops such as pineapples flourished, and growing rice was introduced. Korean, Chinese and Japanese workers were brought to Palau. In 1935, Japan withdrew from the league of Nations, and violated the non-military mandate requirements. By 1939, Palau had become a closed military area, and the islands were being fortified. Seaplane ramps were built, and heavy gun placements overlooked the harbors and reef passages. Palauans were moved away from their villages, and in 1944 most Palauans from Peleliu and Angaur were moved up north to Babeldaob, and had to survive on whatever they could find in the jungle. The Americans bombed Palau heavily from the air in July, and invaded Peleliu and Angaur in September 1944.

Hisakatsu Hidikata (1901-1977)

Hisakatsu came to Palau in 1929, and got a job as an art teacher on Koror. It was his idea to teach the Palauans to carve their stories on small pieces of wood instead of buildings, and sell them to the Japanese on island. This was the start of the storyboards, which are today Palau's most popular souvenir item. Hisakatsu also did a lot of research on the stone carvings and ceramics of Palau, and published several papers on this later. He left Palau in 1931 to spend several years on Satawal and other islands, and visited Palau again in 1938 before returning to Japan. Some of his artwork can be seen in the lobby of the Palau Pacific Resort.

Koror under attack during the American air raids in July, 1944

Former President Ngiratkel Etpison (1925-1997)

During the Japanese occupation there was limited education for the Palauans, they were being taught Japanese language, manual skills, and obedience to authority, with few other opportunities. Local customs and traditions were strongly discouraged. The idea of youth groups was introduced, to provide volunteer labor units for the Japanese Navy. Some of the brightest young men from the youth groups were shipped out to other Japanese controlled countries for further training.

One such Palauan boy was Ngiratkel Etpison. He was sent to Makassar, Celebes, to be trained for shipboard duties and mechanics in the Japanese Navy. In 1944, at 19 years old, he was sent back to Palau on the Japanese military ship Akita Maru, at the exact time that the American air raid operation Desecrate One took place over Palau. The Akita Maru was hit and caught on fire about 5 miles west of Angaur Island. Out of the 17 Palauans on board, 15 survived and swam for days to get to Angaur. Ngiratkel even dragged along his suitcase, which contained all he owned in the world, the most valuable thing being a bottle of liquor he had brought back for his father. After the war he worked himself up from nothing, using old Japanese compressor parts to sell ice candy, and later became one of Palau's prominent businessmen and leaders.

In 1989, while he was President of the US Trust Territory of Palau, he met with a group of American WWII veterans who had come to Palau for their reunion. I still hear him chuckle as he told us about how embarrassed they had been when he had jokingly asked them which one of them had shot at him and his ship. He felt it wasn't his war, that the Palauans were just caught in between. When asked the differences he had seen between the Japanese and the Americans, he told me;" The Japanese treated us like slaves, but taught us how to work hard and use our natural resources. All the Americans did after the war was sent money and canned food, which made our people lazy and dependent on the United States."

Ngiratkel Etpison (*center, bottom row*) on Koror, with his Japanese teachers and fellow students in the 1930s.

Japanese oil tanker Iro on fire in the Rock Islands in 1944.
Sixty years later, the Iro and wrecks of dozens of other Japanese ships sitting on the bottom of the lagoon have become an added attraction for scuba divers. Seen below is the Iro now. It is 470 feet long, and was first hit by a US sub torpedo before being bombed by US planes.

Ship and plane wrecks are scattered all over Palau, most of them in the Rock Island lagoon, where they were shot down by American planes during operation Desecrate One in 1944. Several American planes were shot down by the Japanese also. Some of these wrecks are still missing, and some have been found as recently as this year, mostly by local dive operators and American wreck hunters. Most of the planes that crashed on land were removed after the war, but there are still a few that can be seen, the easiest to get to are those on Angaur Island.

In February 2002 a local dive shop found an American plane which probably still has the remains of the pilot, by accident when recovering an old anchor in ten feet of water in the main harbor. In January 2003 an American team found another American plane they had been searching for, also with crew remains still in the plane in 50 feet of water, untouched since 1944. This B-24 plane had 11 crew, and the 3 survivors were later captured and executed by the Japanese military on Palau.

The wingman of former President Bush was also shot down over the harbor of Palau, and has not been found as of yet. The shipwreck of the destroyer USS Perry, the only American ship to have sunk in Palau in 1944, was also recently found by a local dive operator in 260 feet of water off Angaur. Most divers come for the reefs and fish life, but Palau has some very interesting wreck dives to offer also.

Jake seaplane in shallow water off Airai Channel.

Zeke (Zero) Fighter on shallow reef (Ngaremediu)
Jake twin float seaplane in 40 feet off Ngerekebesang.

Traditional Palau

Palauan legends tell of a fallen giant called Uab, who formed the northern islands. This legend originates from Peleliu and Angaur, where likely the earliest settlements of Palau were. The northern part of Palau was called Babeldaob (Babelthuab) as it still is called today, while the southern areas were called Iouldaob, a name rarely used anymore. Palau was divided into districts, which were always exchanging goods and women, but were also constantly fighting each other for power and standing.

Traditionally, the Palauan men would take care of the warfare, hunting, fishing, trees, making coconut wine, house and canoe building, sailing, and money drilling. The women were in charge of the taro patches and farms, children, weaving, making clay pots and grass skirts, cooking, and tattooing.

Long before European contact, Palau already had a complicated social structure. Each district had governing chiefs, selected from the four most powerful clans. Although both maternal and paternal kinship was acknowledged, the maternal family lines and obligations were more binding and important. Wives and children were basically on loan to a husband, and complicated customs and traditions obligated a man to pay food and moneybeads to his in-laws throughout the marriage. Thus, girl children were preferred, since they were profitable, while boys would only cost a parent money. Sisters and brothers relied on each other for financial support, even after marriage. A lot of these things are still the same way today, although the Palauans are now westernized and educated. The traditional customs related to childbirth, house building and funerals still take place with payments in turtleshell plates and beads, the only difference now the addition of US dollar use.

Each village had meeting houses called Bai, some were reserved for the chiefs and off-limits to women, while others were used as club houses by both men and women. Local wars were fought with spears, clubs, and simple hand weapons until the Europeans introduced guns in the late 1700s. War prisoners were always killed, and an enemy trophy head was put on display in front of the winning chief's house. Villages at times would sacrifice one of their own young men to ask favors from other villages, or as a payment. The election of certain chiefs, or the opening of a new Bai could also require a human head as payment.

Early visitors describe the Palauan men as totally naked, heavily tattooed on the legs, while the women wore grass skirts, and were also tattooed. Both men and women dyed their teeth black and chewed betelnut. Sometime in the 1800s, Palauan men started wearing loincloths, probably introduced from Yap. By 1900, the Palauans started wearing more foreign clothes, and the chiefs started building foreign-style houses.

Sex before marriage was very open and free, but sexual relations between blood relatives were carefully avoided. Interesting situations were created by the existence of Bai girls. Groups of women were exchanged between villages, either as a war payment, voluntary, or kidnapped, and entertained the men in the Bai for months. Marriages often resulted from this, or the women would return to their own village with a payment for their village in the form of moneybeads. Married women were sometimes required to go also, unless their husband could pay off the village chiefs to keep her home. Violence often erupted between the Bai girls and the village women, since they had to feed the girls who were entertaining their men. Children resulting from the Bai girl experience were often adopted by family members if the girl didn't get married. The German administration stopped this practice around 1900. Sailors who had been taking advantage of the Bai girl situation had caused many diseases, and the Palauan population had dropped dramatically due to small pox, influenza and syphilis.

Women had considerable influence over land and money, and were the ones selecting the new chiefs. One's rank in a clan was determined by birth and family background, but could also be increased by personal achievements. Palauans strongly believed in village, nature, and ancestral Spirits. Priests were paid to make offerings and requests to the Spirits. Although most Palauans today are Christians, they are still very superstitious, and local priests are still consulted and paid for services relating to offerings and spells.

Canoes & Rafts

In the past, simple bamboo rafts like the ones at left and on the next page were used to fish in shallow water.

Palau was famous for their canoes, which were used for fishing, transportation, racing and warfare from Kayangel to Peleliu. Only Peleliu and Angaur had heavier ocean-going canoes for cargo (**Dochu**). All Palauan canoes were made out of a single piece of tree trunk. The **Kaeb** sailing canoe had a very light, strongly curved hull like a half-moon, and was used for racing. This was the masterpiece of canoe building according to the Europeans who saw this canoe in the 1800s. The slightly heavier **Kotraol** could be paddled or sailed. In the 1900s the **Brotong**, a heavier version of the Kotraol, was the last canoe type used. The **Kabekel** was the largest war canoe, it was only paddled.

Although some old men still saw canoes in their youth, the craftsmenship has been lost. Recently made canoes do not have the unique deep shape, lightness, and fine shell inlay decorations and details that the Palauan canoes were known for in the past.

Below, left: The late Ilapsis Edeluchel Eungel (at the back) on a Brotong canoe he built. The last canoe that he built in 1998 can be seen in the Etpison Museum on Koror.

Below: The late Imeong Etibek was famous for his detailed miniature canoes. He was still carving at 96 years old.

Open ocean canoe from Angaur Island. Bottom: Canoehouse on Koror, with from left to right: Kaeb sailing canoe, Kotraol sailing canoe, Kabekel paddling canoe.,

Kotraol canoe, photograph taken by Kubary in 1883.
Canoe race between Yap and Palau (right) in Tomil harbor, Yap. The Palauan canoe is a Kaeb racing canoe.

Kotraol

A sailing and paddling canoe. Palauan canoes were always made from a single piece of wood, and the sail and hull could be switched to go either direction.

The type of wood used for making canoe hulls was Breadfruit **Meduu**, the Ironwood **Btaches**, or **Ukall.** For **t**he outrigger pieces of mangrove wood were used, and the float was usually out of Breadfruit. The canoes were decorated with white egg cowries hung on strings, and white shell inlay on the hull.

The sails were woven out of pandanus leaves by the women, and sewn together by the men with coconut rope.

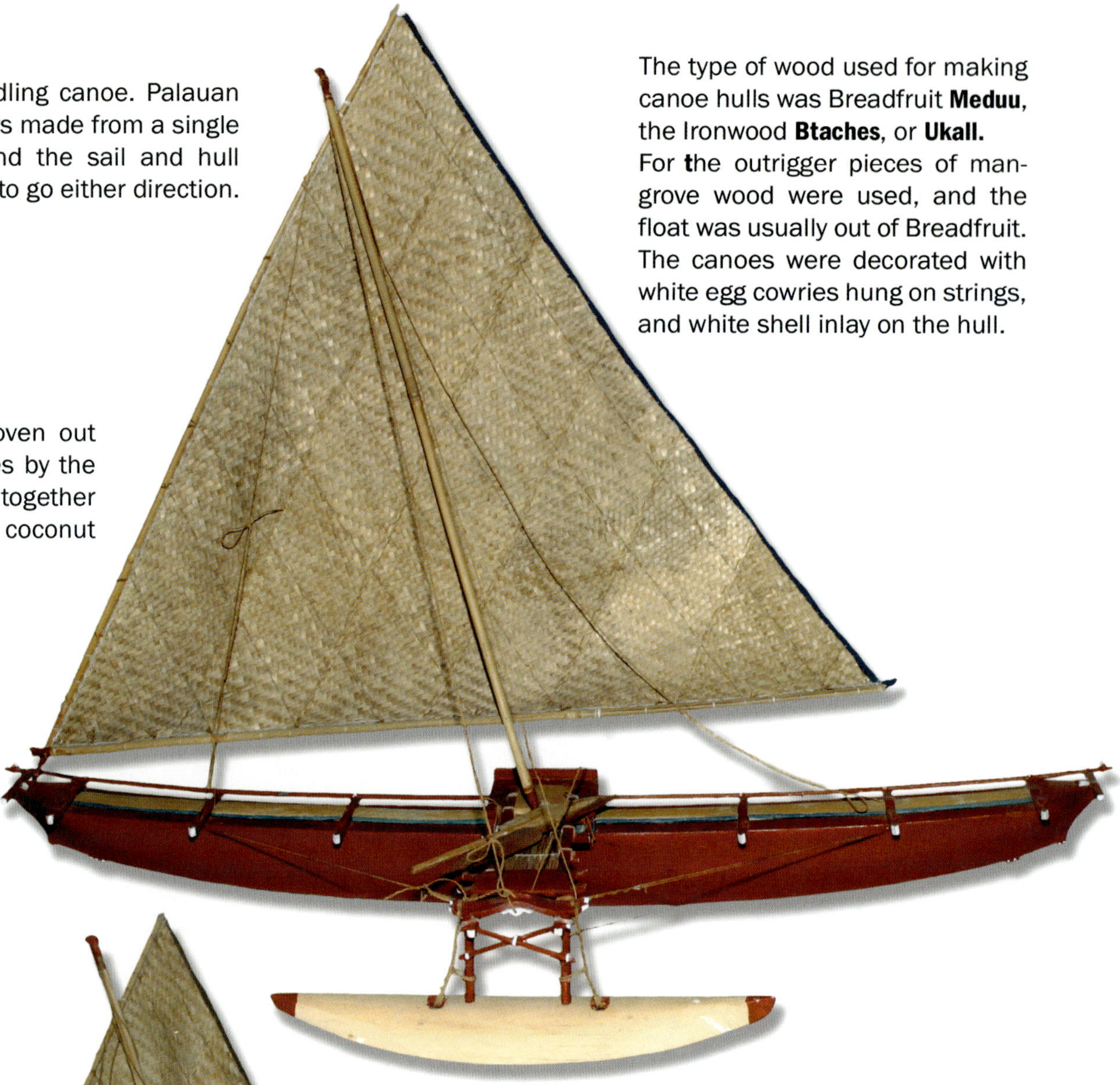

Mlai ra Bul

This canoe was a mix between canoes from the Southwest Islands and Palau. The hull shape was similar to canoes from Yap, but the rigging was done in the Palauan style. This canoe was only used since the mid-1800s.

Kaeb

This was a fast, elegant canoe, used to send messages during war, and there were frequent races between the districts with these canoes. The hull was sharply curved to allow for speed, and always sailed by 4 crew. Two carved kingfisher bird shapes decorated the outer thwart bars, and egg cowries hung off the sides.

Kabekel

This huge paddling canoe was elaborately decorated with carved faces, shell inlay in the shape of animals like birds and fish, and egg cowries hung off the sides. When used for warfare, a loose board in the shape of a canoe hull was placed over the outrigger as seen here, with dozens of egg cowries on the top. This was a warning sign to the enemy. There was also a wooden post, used to hang a trophy head from a dead enemy, which was displayed on the canoe on the way back home. This canoe could be over 50 feet long, and seat up to 32 warriors on top.

Taro

The food of Palauans in the past, before the importation of rice and canned foods, was very healthy, and consisted mostly of seafood, fruits, vegetables and taro tubers. Pigeons, seacows, pigs, and turtles were only eaten on special occasions.

Palauan women maintained and cultivated their individual taro patches, and the condition of their patch was a measure of their standing in the community. Baskets of taro were used as a money value like the dollar, and are still an important food exchange between clans during funerals, births, etc. Today, many older women still have taro patches, but most now use foreign workers to maintain them. The younger generation is less than enthusiastic, and you don't see many young girls going out to the taro patches.

There are two types of taro, the yellow **Brak**, and the gray **Kukau**. Taro is boiled and eaten sliced like a potato, but the chiefs are served Kukau which is boiled and pounded. It is also often boiled, ground and mixed with coconut and sugar as a dessert. The young Kukau stems are peeled and boiled with young leaves, mixed with coconut milk and seafood, and served as a soup.

Blai and Spirit houses

A house is called **Blai** in Palauan. Houses had 2- 6 doors. A house with 6 doors was only allowed for the highest chiefs. During Kubary's time in the 1880s, the only 6-door house of Chief Reklai in Melekeok was burned down in 1882. When Kramer arrived in 1907, Reklai had a new 6-door house, and Ibedul had a 5-door house. Lower chiefs and commoners had to pay moneybeads to their chiefs to be allowed to add an additional door to their house. According to Kramer, it cost 4 **Kluk** and 1 **Kldait** moneybead in 1907 for a 4-door Blai on Koror. All the doors had different names and purposes. The middle door for instance was the seat of the eldest family member, and was where dead bodies were viewed before burial. The inside ceiling beams of the houses of the highest chiefs were often decorated with painted carvings like those of a **Bai** meeting house. Houses close to a Bai were often for priests only. The walls were made out of thatched nipa leaves and hammered bamboo slices, and the door openings could be closed with sliding or hanging woven coconut mats. In some of the areas where the chief was also the priest, you could find the most beautiful houses. In 1907, Keklau still had one 2-story house, which was called a **Sob**. Semper, Kubary and Kramer all describe huge house structures called **Ngasaosech,** which were shaped like a cross, with 4 rooms and carved posts, and which may have been used as priest houses or temples.

The Spirit houses that were built in front of a house were elaborately decorated and painted, and were probably the beginning of the beautiful Bai meeting house structures. There were 5 types of Spirit houses, some were only a carved post, while others were as large as a house. Offerings were often put inside in the form of betel nut or food to the village God, or ancestral Spirits of the individual clans. When I came to Palau in 1985, I still saw two of these Spirit houses being used, but today there are none left out in the public. Even though most of the Palauans are now Christians, the offerings still take place in private, and there are still priests of the local **Modekngei** religion and medicine men and women who cure spells, put curses, make offerings and claim to cure or hurt people mentally and physically with their local herb medicines.

Chief Louch in front of his house on Koror in 1907, with a cooking hut in the back, and Spirit house on the right. He became Ibedul in 1911.

Chief's house on Peleliu with a small Spirit house.
Chief's house in Ngerchelong with a very large Spirit house.

Large clay pots, called **Bekai**, were used as rain water catchments and storage. Smaller wooden vessels and coconut shells were used to scoop out and move the water. The photograph at left shows the way the water was guided into the pot from the tree trunk with leaves. It also shows the woman in front scooping up water, and the woman in the back with a wooden vessel on her head.

The photograph below shows a rare glimpse into the house of Chief Reklai of Melekeok in 1907. Notice all the different shaped wooden plates which were hung from the beams on wooden hangers. Identical hanging racks and the same wooden spoons with carved figurines that were used on Palau in the 1800s, are still found with some tribes in the northern Philippines. Flooring of a Palauan house consisted of bamboo, and a cooking pit is visible on the left side in the back of the room, with a large, probably clay pot for cooking balanced on stones.

The wooden beams on the ceiling were used as storage places for mats, baskets and other things.

This photograph shows food plates laid out in a Bai meeting house during a feast. The wooden food plates had different shapes and names depending on what food they were used for. Certain chiefs, like the Reklai of Melekeok and the Rekemesiik of Ngatpang, were not allowed to eat from any plates other than their own, and had special cooks to prepare their food.

Below you can see women preparing for a village feast, carrying plates and baskets with food on their head. These feasts often took weeks of preparation, and went on for several days also. The German and Japanese administrations strongly discouraged these type of traditions, because they wanted the Palauans to work.

These photographs were taken by Kramer around 1907.

A group of young Palauan girls.

A meeting for women only, taking place at a Bai on Koror in 1907.

traditional wear was simple, the men went naked, and women wore skirts consisting of two pieces, held together with a belt. Daily wear skirts were made from coconut, hibiscus and pandanus leaves for work. For special occasions and feasts, the skirts used were made from the fiber of the trunks of the hibiscus tree **Chermall** *(Hibiscus tiliaceus)* and **Lap** *(Ambroma augusta)*. This fiber was soaked in saltwater, scraped, dried, and painted with clay and turmeric mixed with coconut milk before being sewn together. There were over 12 different types of skirts, and certain clans had the rights to certain decorations and colors. For instance, only 2 clans on Palau were allowed to decorate their skirts or hair with the feathers of the Tropicbird **Dudek**, and only 2 clans could use black skirts.

Palauan women always wore a waist cord, woven from pandanus with a black diamond pattern, or Dugong skin. Later, imported leather belts became popular. Hip belts holding the skirt were made of black and white coconut/ shell rings, and a long time ago Palauans had the Kau belts (left) made from pieces of orange Spondylus shell, which are considered valuable on Yap, and cannot be found on Palau anymore. Today Palauans use mostly simple braided cords, and prefer yarn skirts, which are only used for dances and the 1st Childbirth ceremony.

Carved Trochus bracelets.

Tropicbird feather hair decoration.

Special feasts, or **Mur**, were a favorite pastime of the Palauans in the 1800s, but were stopped by the German administration, around 1900, since they were very time-consuming and the Palauans were not getting any work done. Kramer in 1907 describes 4 types of feasts. The first was a village feast, also called **Mur beluu**, the second was on the occasion of the opening of a new **Bai** meeting house, the third was given by a husband in honor of his wife, and the fourth was to celebrate the re-building of a village destroyed by war. A lot of food preparation and dancing was involved. Women dancers would stay in the village Bai for up to 3 months, and the men were not allowed to enter during that time. They were fed by the village men, who would go out fishing for the feast. The bodies and faces of the women were painted with clay and turmeric stripes during the dancing, which would take place several times a day. A feast could take from 3-7 days. Rows of taro on carved wooden benches, and large containers with syrup would be displayed for show of wealth and to feed the visitors. Invited guests were expected to bring a payment of moneybeads and toluk plates which would help cover the cost of the feast.

Palau's most famous dance, the men's **Ruk** dance, took place only on one of three occasions: The celebration of a newly re-built village, the sickness of a chief, and the opening of a new Bai. The last real **Ruk** dance took place in the late 1800s, and was very impressive according to foreign visitors.

Women's jewelry was simple, and consisted of turtleshell bracelets, hair combs and earrings, and carved trochus shell bracelets as seen on the previous page. The feathers of the Tropicbird were only used as decoration by certain clans.

Both men and women were tattooed. Men heavily on the legs up to the thighs, with other loose designs on their arms and chest. According to early visitors, "Their legs looked like they had been dipped in ink..". Women started getting tattooed during puberty, and the first was a large triangle covering their genital area. They had to have this done before any sexual intercourse could take place. After that, they would start with two fingers, up to above their elbows, before starting on the legs. Designs were mostly lines and triangles, and the amount of tattooing depended on how much they could afford to pay the woman in charge of doing it. So high-ranking rich women and men had more tattoos. The tattooing was stopped around 1900, and the only women still alive today with these tattoos are over a hundred years old.

Female tattooing *(after Kramer, 1907)*

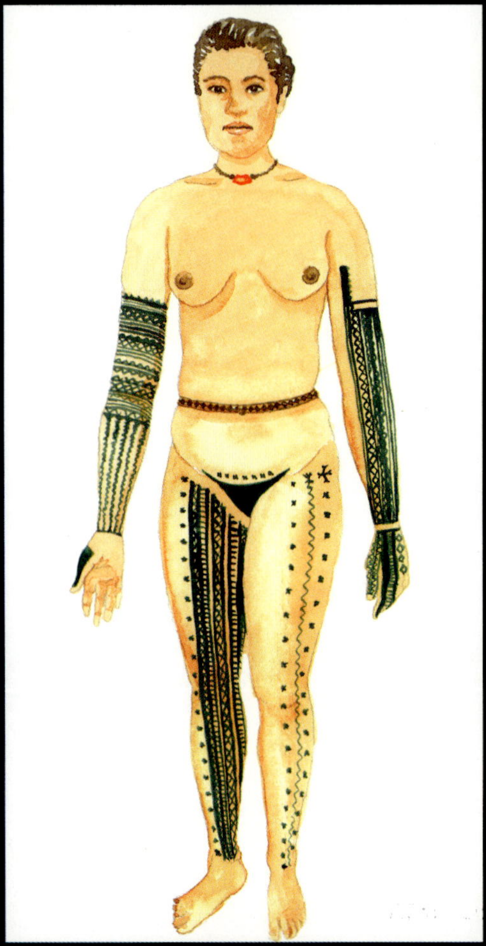

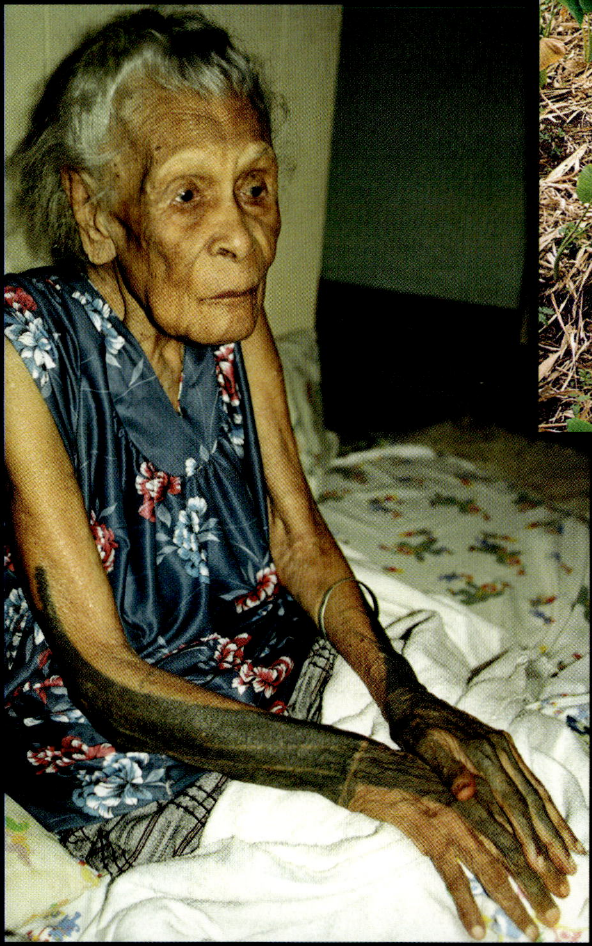

Taldil Olkeriil, over 100 years old. There were no birth certificates yet when she was born.

Women dancers took place on a bamboo platform with a slanted thatched roof, called **Diangel** like a canoehut. There were many different type of dances. Unlike the spectacular Polynesian dances, the emphasis in Palau during the dances was on the often funny or erotic meaning of the chants, not so much on the movements.

Wood

Considering the very simple tools and shell-blade adzes used, the craftsmanship of the Palauan carvers in the 1700s and 1800s was amazing in detail and design. The delicate shell inlay would be difficult to copy even with today's tools. Fortunately for Palau, the early European visitors brought many items from Palau back home with them, especially the German researchers Semper, Kubary and Kramer. Even the items brought back by Captain Wilson in 1783 are still preserved in the British Museum today, and are the oldest surviving Palauan artifacts. Many of the items preserved in Europe cannot be found on Palau now, and have not been made or used since the early 1900s. A lot of young Palauans have never seen these items, since the humidity and termites destroyed most of the wooden heirloom items on Palau in the last century. If it wasn't for these foreign collections, a lot of the best examples of Palauan woodcarving would have been lost to the new generation of Palauans today, who grew up with imported plastic and metal. I am including photographs of some of the European collections here, in the hope to inspire Palauan carvers of today to go back to their roots.

All the wooden plates that would come into contact with liquids, were painted with clay, and varnished with the oil of the *Parinarium* nut **Cheritm.** The meat of the nut was boiled for days, and then the oil was squeezed out. This oil was only good when old and cold, so it was kept for months under water in the taro patches. The plates were then polished with bamboo leaves, and another coat of oil would be applied, after which each plate was rubbed with a mix of coconut milk and turmeric.

Buk plates were used for serving starchy food, and special ones inlaid with clamshell designs on the edges were called **Madelchedeng**. Raw fish was served on **Chongbal** platters, while cooked fish and birds were served on the large, high **Ongall** plates.

Left: A coconut spoon with wooden handle. Similar spoons were used and can still be found with hill tribes in the northern Philippines.

Below: This wooden weapon was collected by Kubary in the 1880s, and was probably a showpiece belonging to a chief or priest. The shape is that of a stylized bird.

Both items: Staatliche Museen zu Berlin, Preuβischer Kulturbesitz-Museum fur Volkerkunde.

Inlaid serving plate for special occasions. 1783, British Museum collection

Serving plate out of a single piece of wood. 1910, Etnison Museum collection

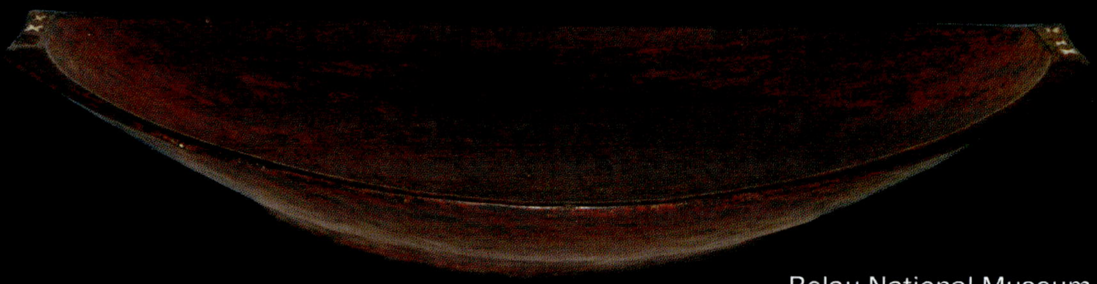

Belau National Museum collection

Belau National Museum collection

Belau National Museum collection

Etpison Museum collection

Etpison Museum collection

Food serving plates with the clamshell- shaped shell inlay characteristic of Palauan plates.

Medicine bowl for “love potion”, Around 1900.
Etpison Museum collection.

Taro pounder and plate. Around 1900. Etpison Museum collection.

Offering stool for taro, used during feasts to make offerings to the Spirits, also called “**Toluk ra Rubak**”. Shell inlay with spider design. Staatliche Museen zu Berlin-Preuβischer Kulturbesitz, Museum fur Volkerkunde.

Taro display stand for village feasts. 1862, Museum fur Volkerkunde, Dresden.

Offering stool for taro with spider shell inlay. 1783, British Museum collection.

Taro stool with shell inlay, 1862, Museum fur Volkerkunde, Dresden.

Plate with shell inlay and pot in the shape of an owl, with glass bead eyes.
Staatliche Museen zu Berlin- Preuβischer Kulturbesitz, Museum fur Volkerkunde

Large standing pot to hold drinks during village feasts, called **Ilengel**.
Waterpot and stack of food platters. 1800s, British Museum collection.

Large wooden pot to hold sweet drinks for chiefs. 1783, British Museum collection.
Carved in the shape of a Megapode **Bekai** bird.

Large wooden pot carved in the shape of a duck or whimbrel bird.
1899, Museum fur Volkerkunde, Dresden.

Wood inlaid jar, used for storing the antique moneybeads. 1783, British Museum collection.

Another example of a wooden jar with intricate shell design, 1783, British Museum collection.

Hanging pot with shell inlay, 1783, British Museum collection.

Inlaid pot, inside painted with yellow clay. 1800s, Etpison Museum collection.

Serving plate, Belau National Museum collection.

Hanging pot with inlaid face, collected by Karl Semper in 1862 - Museum fur Volkerkunde, Dresden.

Bai

A traditional meeting house on Palau is called a Bai. Every village in the past had at least one of these meeting houses. Considering that only primitive clam-shell adzes were used for the construction, these were real masterpieces of architecture. The first thing foreigners noticed about the Bais were the elaborate, often sexually explicit decorations and stories, both on the inside and outside of the structures. Recording their legends and stories in the paintings was the Palauan way of preserving their historic events, since there was no written history. A Bai was always built on top of a stone platform, which often also served as a meeting place for the Chiefs, who sat against four upright stones on the corners of the platform. The four corner posts of the Bai represent the four leading Clans of a village, whose Chiefs rule the area.

The Chief Bais were the most elaborately decorated Bais, used for important meetings, and off limits to women. The other Bais were used as clubhouses for young men, who learned about fishing, hunting,carpentry, and sex in these buildings. Bai girls who were brought in from other villages to entertain the men in these clubhouses were fed by the local women, and received moneybeads upon return to their village after several months. This was an important income for villages. Villages ordered Bais from other villages, and a Bai was paid for with moneybeads per finished sections.

Left: Detail of one of the Bais on Koror in the early 1950s.

On rare occasions, two-story structures were built, both as houses (**Blai**) and **Bais**. The last known two-story Bai in Airai was photographed by the German researcher Kubary in 1883, right, the only record now of such structures ever existing, since Palauans today do not even remember. According to Kubary, these structures were not very practical, and built more to impress.

When Kramer visited Palau in 1907, this same Bai, with the name **Chotang**, was already damaged, as you can see on the photo below, right, where it stood next to two other Bais in Airai.

Two Chief Bais in Ngerchelong with a Spirit house. All these photographs were taken in 1907 by Kramer.
Two of the three Bais in Meketii, Koror.

A Bai and two Chief Bais in Ngeremlengui.
The three famous Chief Bais of Melekeok.

A female carving called **Dilukai**, as seen on this photograph at left, was hung above the entrance of many Bais in the past. Some say it was a sign against incest, others believe the carving represented the tradition of the **Bai girls**, women from other villages who would live in the Bais for several months to entertain the men in return for money-beads. Smaller male carvings were hung off the Dilukai's feet. None of these carvings are found on Palau anymore. Four Dilukai are still in European and American museum collections.

Below: Young boys learned the art of constructing a Bai and all the individual parts involved, by practising on miniature Bai structures such as these shown here. A village would normally order a Bai from another village, and pay for it with beads, since Palau's social system was based on alliances and exchanges. A Bai was like a puzzle, built in separate pieces, and taken apart to be moved to the village that ordered it. The stories on the Bai were from the area, and represented famous Spirits, local heroes, warfare and a lot of sexually explicit stories. The stories were outlined with light carvings, and colored in with yellow, red and black clay, soot and lime, sealed with the oil of a local nut.
Favorite symbols included roosters, carved faces, fish, clams, and the money symbol, a black circle with a cross in it.

Original Palauan Dilukai from the late 1800s. These cannot be found on Palau anymore.
Staatliche Museen zu Berlin- Preuβischer Kulturbesitz, Museum fur Volkerkunde.

Beam of a Bai meetinghouse, late 1800s.
A half size Bai model on display in Berlin, Germany.
Staatliche Museen zu Berlin- Preuβischer Kulturbesitz, Museum fur Volkerkunde.

This Chief Bai from Ngeremlengui is the only Bai known to have its front decorated with only one symbol, that of a spider. The spider was connected with the childbirth ceremony, so that story may have originated from this area, or the Bai may have been used in connection with childbirth.

Some of the Bai stood over a century, with regular maintenance from the villagers. The Ngermid Bai on Koror where Captain Wilson stayed in 1873, (see page 126) was still standing when Kramer visited Palau in 1907. When a Bai became too shaky, it would be supported by wooden poles on the sides, like you see on the photo below, which is a Bai from Mengellang in Ngerchelong. All of these photographs were taken around 1907, when still over a hundred Bais were found in villages on Palau. Thanks to the efforts of German researcher Augustin Kramer and his wife, photos and paintings of all these unique and culturally important Bais were preserved for us to enjoy today. Most of these Bais were destroyed by the 1940s and were never rebuilt after WWII. Whole villages had been moved away from their traditional areas by the Japanese, and only 5,000 Palauans were left alive in 1945, compared with an estimated population of over 40,000 in the 1700s.

Front and back of the Airai Bai with local dancers. This is the only Bai that survived from the 1800s, but today it is in desperate need of repairs.

Melekeok was famous for the best Bais on Palau in the past, but most were destroyed by British warships in the 1800s, and later by typhoons. The last Bai was built in 1938, but was soon destroyed.

In 1992, Melekeok opened a new Bai, built on the original stone platforms where once three Chief Bais stood. The opening of the Bai coincided with the inauguration of the new Paramount Chief Reklai Rafael Ngirmang, seen standing in front of the Bai on the next page. The Reklai is the only Chief on Palau to wear a women's skirt and moneybeads during his inauguration ceremony. This was only the second such ceremony in the 20th century.

Below and left: views of the new Melekeok Bai from the air and from the side.

Although the detailed carvings and Dilukais are missing from the new Bais, the recent construction of Bais shows that this unique tradition does not have to die out, and can still be revived. It would be great if each state could build at least one state Bai, not only as a tourist attraction, but to teach their children pride in the traditions of their past.

The Bai at the Belau National Museum. This Bai was constructed in 1990.

Aimeliik State constructed a Bai in 1995.

1st Childbirth ceremony

A series of ceremonies take place in Palau during a woman's first pregnancy, and after the child is born. These ceremonies also form a bond between the father's and mother's clan, and a series of exchanges of food and money take place. It is also a way for the woman and her baby to be accepted into the father's clan.

It is said that centuries ago, natural childbirth was unknown in Palau. When contractions started, her stomach would be cut open with a bamboo knife to deliver the baby, which usually proved to be fatal for the mother. A foreign God in the form of a spider married a Palauan girl from Ngiwal, and delivered his baby the natural way. Ever since, the celebration of the firstborn has taken place on Palau.

When a woman is pregnant for the first time, she would return to her mother's house, and a medicine woman is chosen to be in charge of the bathing ceremonies. Today, the woman usually returns to her mother's house after the birth, just for the bathing ceremonies, which are still practised by most Palauans today.

In the old days, the use of medicinal herbs would start during pregnancy, and continue till afterwards. The girl would be separated from her husband for over a year. This would give her body time to heal from the pregnancy. The husband could come and ask his mother-in-law for permission to take a girlfriend during this period, which was usually allowed, but he was expected to make payments and take back his wife after the ceremonies were completed. This was a common practice in previous centuries.

Today, the bathing ceremonies take place at the mother's house during a 3-10 days period only, depending on the rank of the parents' clan, as soon as the baby's navel has healed.

Left: Leah Asanuma at her Ngasech ceremony.

There are three main events involved in the bathing process. The whole process is called **Omesurech**, and starts with the daily bathing ritual. After that is the final treatment **Omengat**, which takes place in a special hut, Then the public **Ngasech** ceremony finishes it off, when the new mom shows herself and her baby to her in-laws and relatives, who dance and chant around her, and go home with large baskets of food. The Ngasech ceremony usually takes place on a Saturday or Sunday in front of the house, around lunch time. Permission should always be asked from the girl or her parents before taking any photographs of such a ceremony. The Ngasech is done pretty much in the same way all over Palau, except on Angaur Island, where they construct a high wooden seat with steps for the new mother and two relatives to climb up and be seated, and a special food offering is made to her.

In Palau a husband only borrows his wife from her clan, and he pays for this privilege continuously during the time of the marriage. Parents in the past preferred to have girls, because they would bring money into the family, while boys only cost money.

While today a lot of couples marry in church and in court, there is a traditional payment for marriage in the form of moneybeads, which is sometimes combined with and paid for at the time of the childbirth ceremony. The father's family also makes payments to the new mother's family during pregnancy, and after the birth there are two payments; one for the baby, and one for the damage to the mother's body. Payments are made in US dollars, together with moneybeads and turtleshell plates. Afterwards, the money and turtleshell plates are distributed to the relatives of the new mother who helped and prepared food, and to the medicine woman in charge of the bathing.

Because of the old legend mentioned on the previous page, spiders are often associated with childbirth in Palau. The spider referred to in the legend is the huge Orb-web spider **Mengidab Rutkoel** *(Nephila sp.)* The males of this species are tiny, while the females can be as big as your hand. Their silk is the strongest natural fiber known on earth, and was used by early Palauans to make fishing lures.

Right: Carol Tellei at her Ngasech ceremony.

The medicine woman, also called **Mesurech**, starts gathering herbs with the assistance of the girl's female relatives. Every Mesurech has her own secret ingredients and herbs, but some common herbs are used by everyone. One of the most important ingredients used by every clan are wax-apple **Rebotel** leaves (*Eugenia javanica)*, below right. The leaves are boiled and the juice is used to splash on the girl during the daily bathings. Fresh leaves are used to rub her skin with, and should be perfect so as not to scratch her. A medicinal tea is also made with these leaves for the new mother to drink.

Another important ingredient is the turmeric Ginger root **Reng** *(Curcuma domestica)* as seen above, and below left. Palau has a small yellow variety, while the larger yellow and orange ground roots are brought in from Hawaii and the outer islands of Yap. The root is ground, boiled with coconut oil, and drained the night before the bathing starts. The resulting oil is used on the skin, while the drained root is wrapped in fabric to dab on the skin at the end of the bathing each day. By the end of the week everything and everyone involved will be stained yellow.

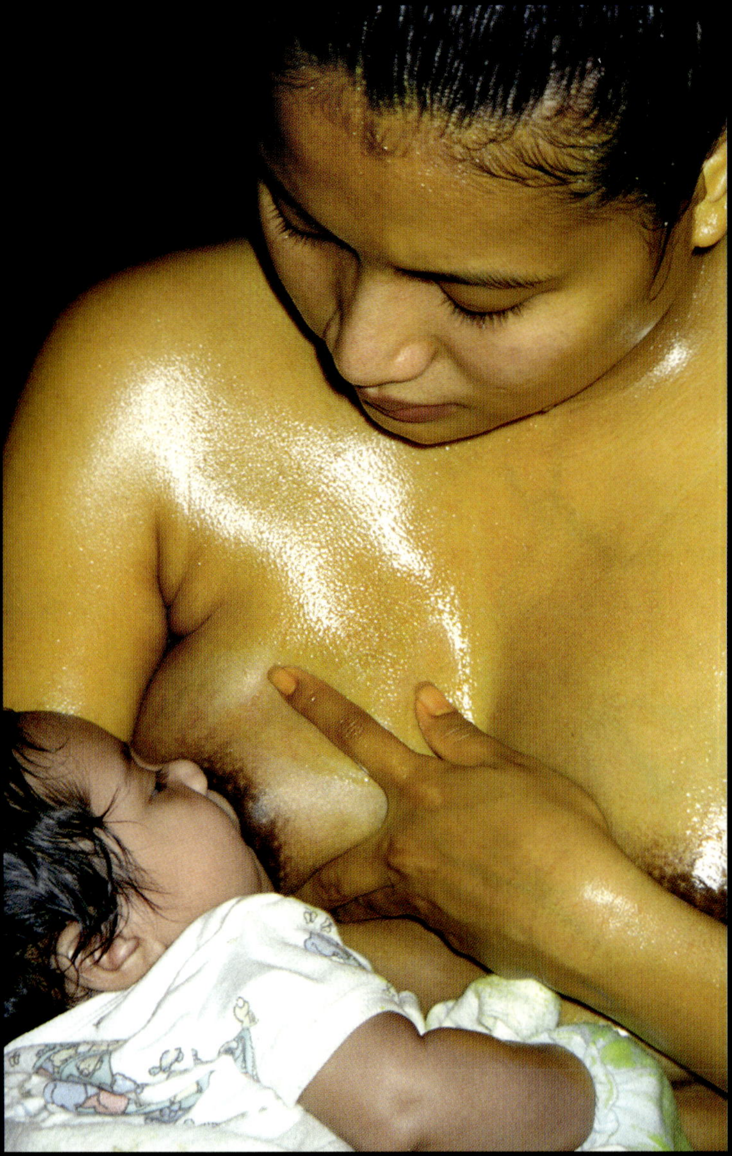

Omesurech

After the baby's bellybutton heals, the bathing period, called Omesurech, starts. Today, it lasts 3-10 days, depending on the rank of the parents' clan. The male relatives go fishing before it starts to provide food for the ladies attending, who will be fixing tapioca for everyone there. The new mother is confined to a small steam hut (**Omiachel**) and is only allowed out to breast-feed her baby. She is not allowed to see her husband.

Several times in the morning and afternoon, she is splashed with boiled medicinal water. Her skin is protected from the hot water by a mix of coconut oil and turmeric called **Kesol**. Local lemon is applied to any stretch marks. In the past, this was done daily for months, so the medicinal healing effect must have been much better.

There are also medicinal teas used for different problems, and herbs to stop excessive bleeding.

After the last day of bathing, a mixture of four raw herbs is chewed to make the uterus contract back to it's normal size fast. This is done while showering in cold water or sitting in a stream, since the herbs make you feel on fire, and taste terrible.

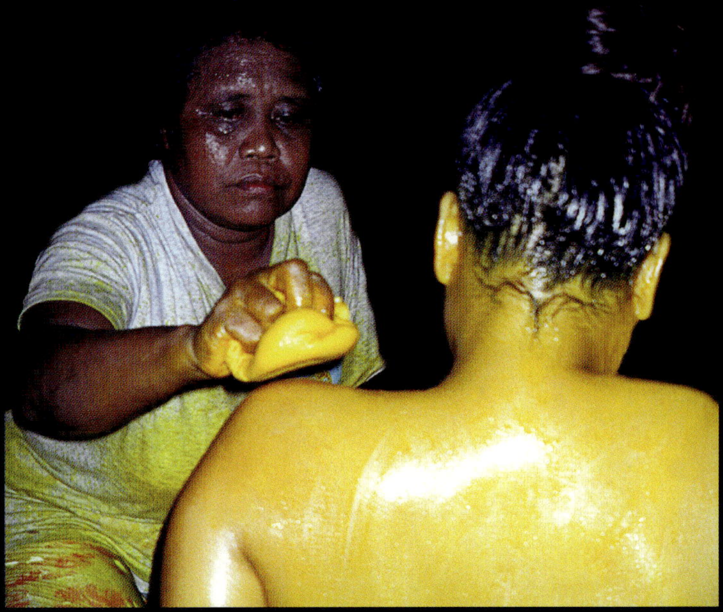

During the Omesurech bathing week, the new mother is slapped, rubbed and wiped with the wax-apple leaves, while a half coconut shell is used to throw steaming hot medicinal water at her from every angle. At the end of each day, her skin is dabbed with Ginger root, to sooth the skin till the next day of torture begins.

Aunties and sisters sit outside the hut preparing food and exchanging stories about their own experiences with childbirth, medicines, men, and life in general.

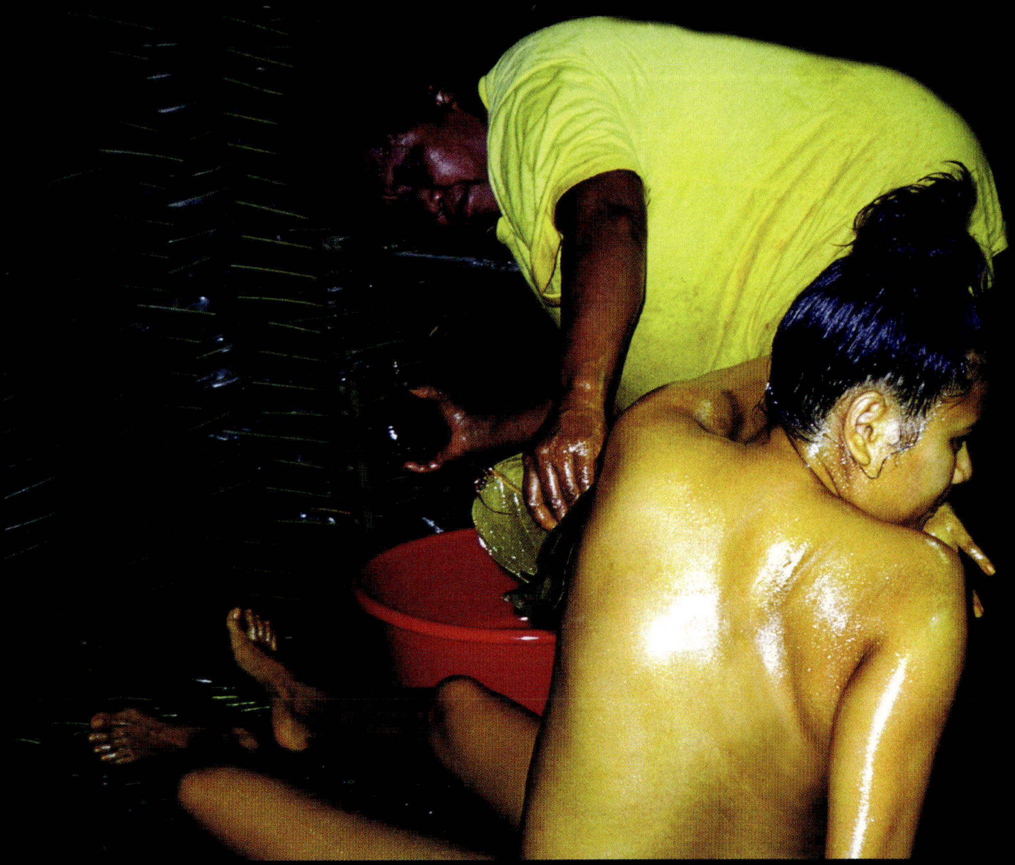

Strings of ground and boiled tapioca wrapped in leaves are hung around the steam hut area, and always more seem to be in the making. Betelnuts are collected and prepared in packages together with leaves and lime. Fish is cleaned and prepared by male relatives, and brought to feed the aunties. Usually, only aunties, sisters and the mother of the new mom are allowed around the steam hut by the **Mesurech** medicine woman in charge.

Meanwhile, the aunties and female relatives are being fed, and enjoy their betelnut. Coconut leaves are woven into mats for the final day, when they will be used by the girl to walk on when coming outside.

Omengat

The final treatment is in a specially built steam hut, called a **Bliukel**. After one last set of wax-apple splashing, the new mom is joined inside the hut by some close friends or relatives who have already had a baby themselves, and who will join her naked for some of the steaming. As soon as the herbs boil, some old coconut shells are burned and mixed with the herbs to create a strong medicinal steam. The girl squats on a toilet-like seat (**Osekuakl**), under which the herb mix is placed on a leaf, to clean out her insides. Bowls of steaming water with herbs are placed inside later also, both to perfume her body and for medicinal purposes.

The new mom is now as good as new, and ready to show herself to her in-laws. She will start getting ready for the final public **Ngasech** ceremony.

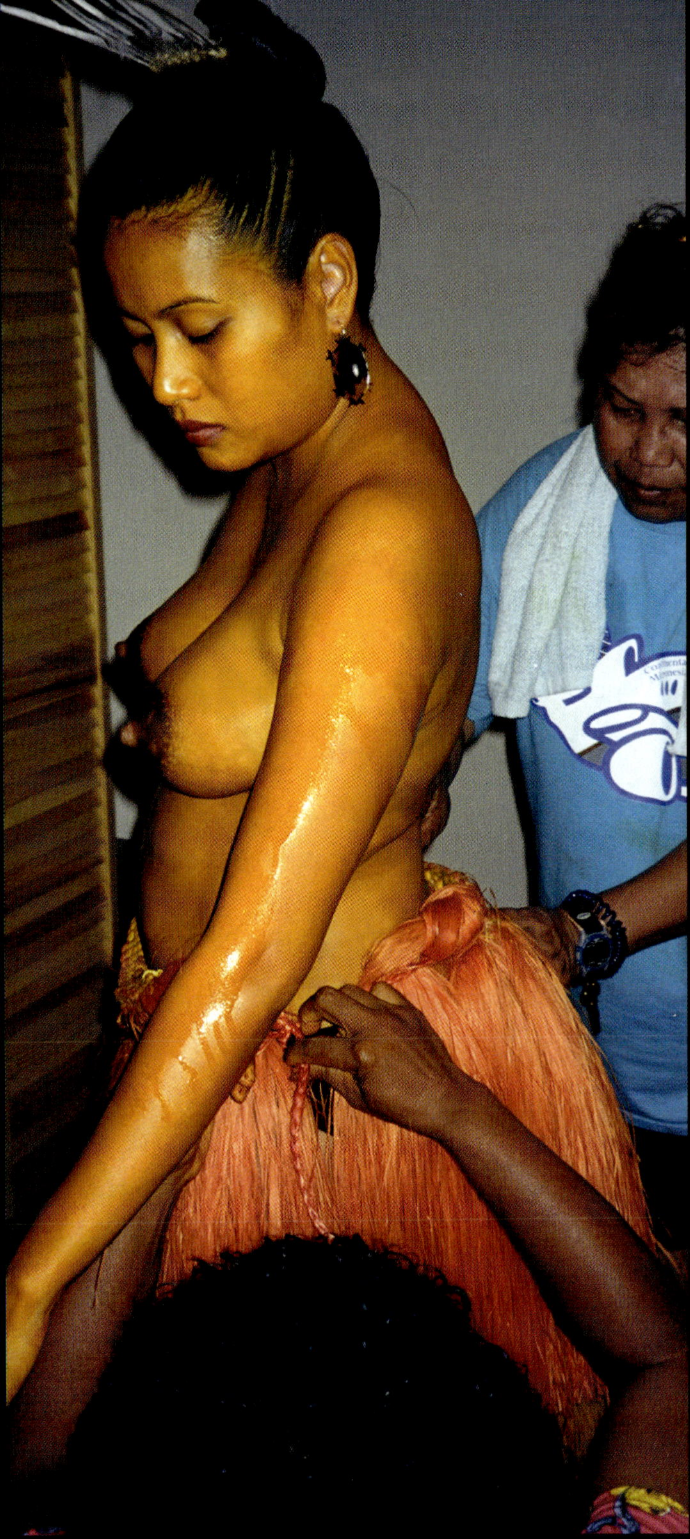

After the final steam bath the girl gets dressed up in a traditional skirt with colors and hair decorations from her mother's clan, and prepares for the final **Ngasech** ceremony. They dab turmeric mixed with coconut oil on her skin, and she then has to wait for her in-laws to show up. They will come with gifts for the baby, and to put moneybeads on her neck for her to wear during the ceremony. Her in-laws will meet with her parents or relatives, and moneybeads, turtleshell plates, and US dollars will be given as payment for the baby and for "the damage " to the mother's body from pregnancy. Sometimes this **Ngasech** ceremony is combined with a marriage payment also if that did not happen yet.

It is the responsibility of the baby's father and his relatives to come up with the payments. In return they receive lunch and food baskets on the day of the ceremony, prepared by the girl's relatives. The payment of US dollars and turtleshell plates will be distributed among the girl's relatives who helped during the preparations and provided food. The medicine lady **Mesurech** will also receive a separate payment for her services.

After her in-laws have put the beads on her, she is ready to present herself and her baby to all the relatives. Mats of woven coconut leaves, called **Telutau**, are laid in front of her while she walks slowly, supported by the **Mesurech** to an area where everyone can see her. The husband's relatives will come forward and chant and dance around her. She will have a **Rebotel** leaf in her hand to wipe the oil from her eyes, and one arm supporting her breasts and elbow. A bowl with medicinal leaves is placed by her feet, and splashed around during the dancing. Flower bouquets, dollar bills, and live bands are unfortunate recent additions to the ceremony. Although the medicinal healing effect is probably less now because of the shorter time period, it is a unique tradition which will hopefully continue to take place for many more years to come.

The moneybead payment is presented by the new father's relatives on two turtleshell plates, and covered with a leaf. Loose beads always have cotton in them to make them easier to handle. The beads are then carefully examined by the girl's relatives, and the payment announced to the rest of the relatives afterwards.

Ngasech

Lunch is served outside to all the relatives from both sides, either buffet style or wrapped. Dozens of baskets of food are waiting to be taken home by the father's relatives.

The girl is brought out by an auntie or the Mesurech, and walks on coconut mats, showing herself to the relatives. Aunties and sisters of the new father will come to her and chant and dance in front of her. Later, the baby is brought out also, and family pictures are taken before the new mother can finally relax.

Turtle shell

In the early 1800s, sea turtle was only eaten as an offering to the Gods and during sickness, and was the highest prized food on Palau, one turtle could cost up to one **Kluk** moneybead. The shell of only one species of sea turtle, the Hawksbill, locally called **Ngasech**, is used to make jewelry and traditional money plates from. The shell of this turtle is thicker and nicer in color than that of other turtle species, and was known in other countries as tortoise shell (even though a tortoise is a land turtle..), and often used for inlay and jewelry in Europe also. The Hawksbill is now an endangered species worldwide.

The plates made from turtle shell are called **Toluk.** In previous centuries they were also used as food plates for certain chiefs, but today they are only used as a women's money. They are exchanged between clans during traditional customs such as funerals and childbirth, and used as a payment for services rendered to women who prepared food during those customs. The plates are also used to present the moneybead payments on. The plates are made by heating the shells so they become soft, and shaping them between two wooden forms. Recently made plates are not nearly as perfectly shaped as the antique ones, and often have press marks, uneven edges. Most of the plates made before 1900 only have two decorated edges on the two ends, while modern plates also have these on the sides of the plate.

Although the Hawksbill turle is highly endangered on Palau, attempts to protect it by law have failed so far, there is a hunting season, but it is not being enforced, you will still see jewelry for sale in the stores, and new **Toluk** plates are still being made. Older Palauan women do not want to register their plates, or for other Palauans to find out how many plates they own.

Turtleshell spoons were often worn on a string with glass beads around the neck or arm, and considered one of the most valuable possessions of a woman, only used for special occasions. Other items made out of turtleshell included earrings, hair combs, bracelets, medicine bowls, small knives used for lime and betelnut, and the special heavy arm rings as shown on the next page, which were only worn by high clan women.

Excavated spoon from Airai, found together with these loose beads.
Etpison Museum collection.

1783 spoon, British Museum collection.

Spoon with beads from Ngeremlengui, heirloom item.
Etpison Museum collection.

Thick medicine bowl from Ngeremlengui, heirloom item.
Etpison Museum collection.

Bracelet made from dozens of turtleshell rings. In the past worn by high-clan women.
Belau National Museum collection.

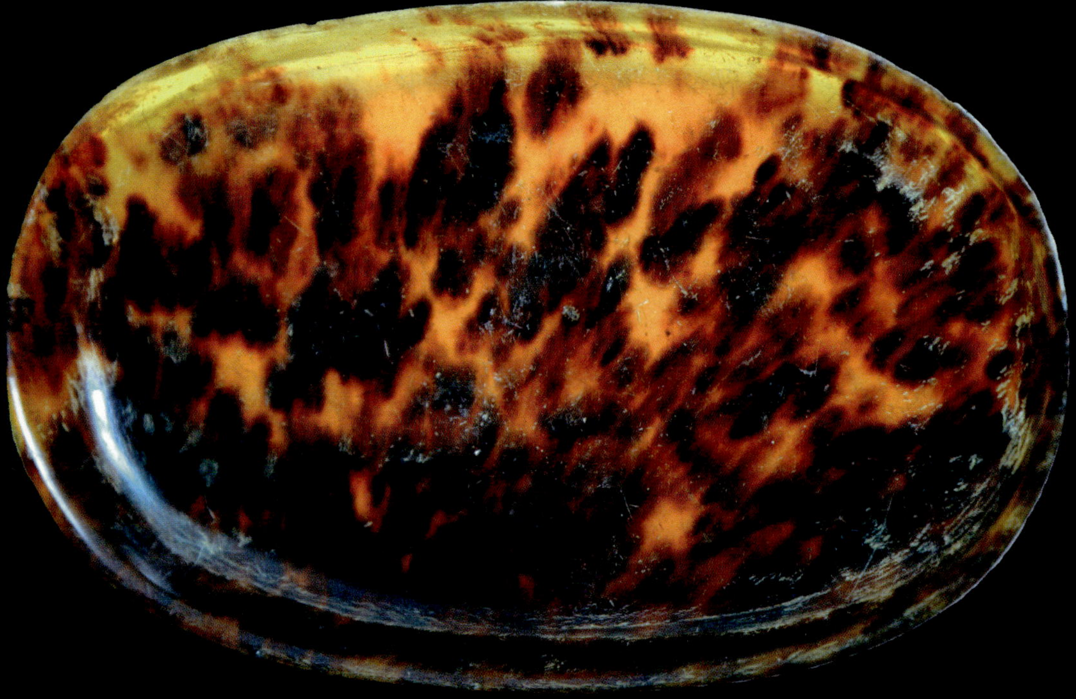

Toluk from 1880. Heirloom item.

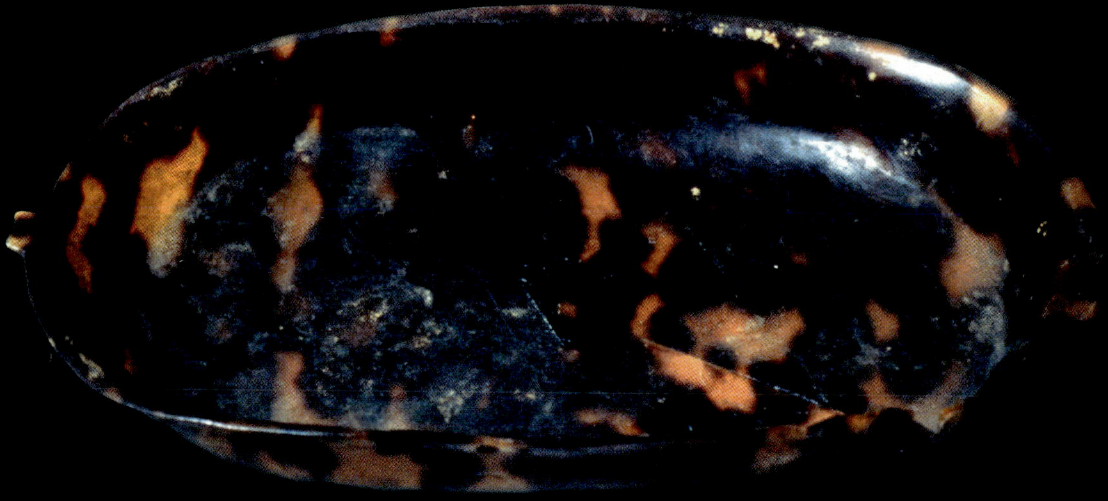

Toluk from 1783. British Museum collection.

Traditional turtleshell plates are called **Toluk.** In the past, these plates were also used as dinnerware for chiefs. Today, they are still used as women's money, and exchanged between clans during funerals, child-birth and other traditional customs. The plates are valued on the size, thickness, as well as the color. Real old plates from the 1800s have two small holes on the back for hanging. The older plates are usually more squarish and perfect in shape, and have only two, or no decorated edges, while the modern plates have six.

Right: My mother-in-law, Yuang Etpison, with her collection of turtleshell **Toluk** plates. Palauan women usually wrap their collections in fabric and plastic, and hide them away in boxes, with little notes mentioning when they received them and for what custom.

Toluk plates lined up for payment distribution during a funeral custom.

Olecholl

Olecholl is a traditionally valuable bracelet on Palau, made from the vertebrae of a Dugong *(Dugong dugon)*, locally called Mesekiu, also called a sea cow. Dugongs are similar in appearance to Manatees, but have a fluked tail like that of a whale. They can weigh up to 375 pounds.

Traditionally, only certain villages were allowed to hunt and eat the meat of Dugongs, and use of the 3 bones used to make bracelets was reserved for the highest chiefs, providing the animals with some protection against over-hunting. When a new chief was elected, he would be allowed to wear one of these bone bracelets. Today, only a few Palauan men wear one as jewelry. They are mostly kept as heirloom items, and have lost their traditional value and use. This is a good thing, since the Palau Dugongs are now a highly endangered species, hunted close to extinction in the last century. On rare occasions they can still be seen around the Lighthouse Channel of Koror, where they feed on sea grasses.

To fit the bracelet on an adult man was a painful ordeal. The inside of the bone could be scraped slightly, but never cut, and the result of putting one on a new chief was often a broken finger or hand. One the next page you can read the earliest observations by foreigners on the bracelets and their use in Palauan society then.

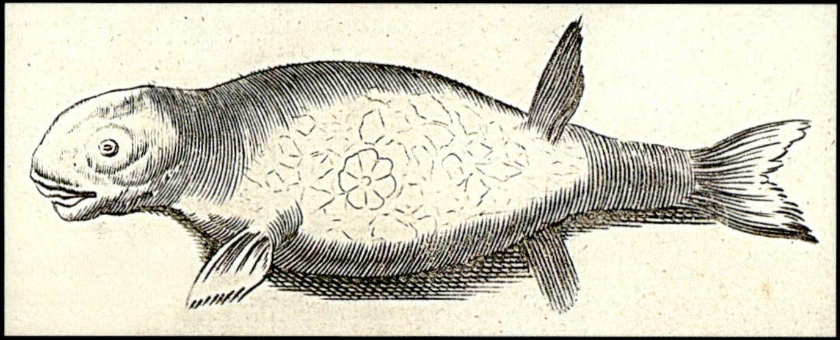

An early European drawing of a Dugong.

Chief Reklai Soilokol with his wife Bilung, wearing an Olecholl bracelet around 1907.

The original bracelet that was given to Captain Wilson by Chief Ibedul in 1783, today still in the British Museum collection.

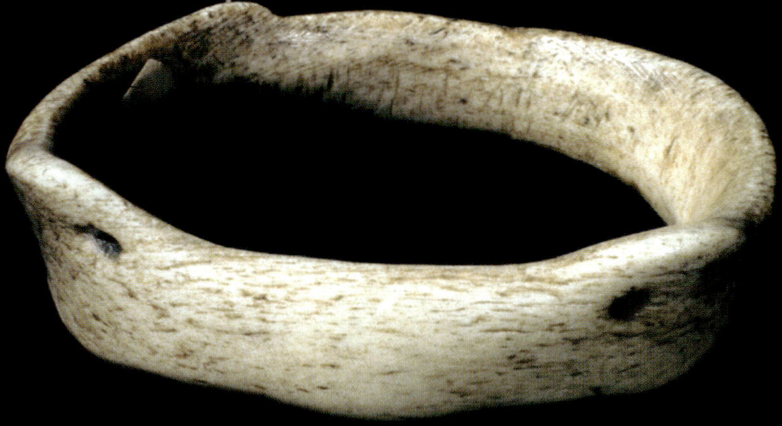

September 1783

Among the prisoners (of Meleok) was a rubak, who had a bone on his wrist; the Pelew (Koror) people, after he was captured, strove all in their power to take from him this mark of dignity, the chief defended it with similar courage, and lost his life in endeavouring to save it.

From: An Account of the Pelew Islands, fifth edition. 1803- George Keate.

November 1783

The King (Chief Ibedul) informed Captain Wilson that it was his intention to invest him with the Order of the Bone, and make him in form a rubak of the first rank. The bone was applied to see if it was large enough for his hand to slip through; being not found sufficiently so, it was rasped away. Raa Kook made a string fast to each of the fingers of the Captain's left hand, and then lubricating the hand with oil, he placed himself behind the Captain, holding him fast by the shoulders; then passing the different strings through the bone, and giving them to another rubak, they endeavoured to draw his hand through; Raa Kook at the same time with his own hand compressing that of the Captain into the smallest compass he possibly could, so that the bone might pass over the joints. The point being at last obtained, and the hand fairly passed through, the whole assembly expressed great joy. Abba Thulle (Chief Ibedul) then addressing Captain Wilson, told him, that the bone should be rubbed bright every day, and preserved as a testimony of the rank he held amongst them; that this mark of dignity must, on every occasion, be defended valiantly, nor suffered to be torn from his arm, but with the loss of life.

From: An Account of the Pelew Islands, fifth edition. 1803- George Keate.

June 1791

One their passage they disturbed a sea monster, called by the natives a Mesekiu, and, from the description they give of it, resembles in many particulars the Mermen of Norway. From this monster they get the bone that invests the wearer with the order of rubak. The account of this fish is so extraordinary, that I shall give it. He said the largest he had seen was sixteen feet in length, and twelve feet in circumference, being as much as two men could encompass with their arms extended; its head is like that of a seal, with arms or fins, with which it feeds itself; the female has breasts with four nipples, just below the arms, where it suckles its young; it has no hair upon its body, being smooth like a porpoise; its tail is in the form of an equilateral triangle; its principal food is beech de mer (sea cucumbers) and sea weed; the male and female are readily distinguished.

From each fish they get three bones, which are of considerable value; the first is cut from the skull, or front of the head, and is of the lowest order; the second is taken from the middle of the head, and is only worn by the Chief rubaks; it is the same as represented in Mr.Keate's Account of the Pelew Islands; the third bone, is that which joins the head and neck, and is the largest; it is worn by men who are not rubaks, but who have signalized themselves by some act of personal bravery.

From: A Supplement to the Account of the Pelew Islands. 1803- John Pearce Hockin.

Udoud

Rudimch collection,
Worn by Erico Rudimch.

A necklace with a whole family collection of moneybeads like this is called a **lek**, and is only shown on special occasions.

Antique Palauan moneybeads, of unknown origin and made from glass and glass paste mixed with minerals, have been and still are used in an exchange system unique to these islands. Today, in combination with American dollars, they are still used for all important events, such as childbirths, funerals, marriage, building of a house, and selection of a chief. Although the Germans introduced the Mark, and the Japanese the Yen, followed by the American dollar, the traditional use of moneybeads has survived. In daily life, a woman may wear a single piece of money, always on a simple black string. The money usually belongs to her husband's family or her parents. Wearing a whole necklace full of beads, belonging to a whole clan, was reserved for special occasions in the past, and the beads were normally carefully hidden or buried. Wars were fought over these beads in the past, and today lawsuits have been filed over the ownership of these beads.

The Palauan moneybead system was a frozen one for a long time, since most of the bead types used are centuries old and not being made anymore today. Occasionally a few beads were brought from Yap in the past century, excavated during farming. There is no accurate inventory of the number of Palauan moneybeads because of the secrecy surrounding them. Owners, especially older people, are reluctant to show their beads, because relatives may scheme to obtain some from them through traditional customs, plus they do not want other clans to know how many beads they own. During the last 15 years, "new" beads have been brought in from the Philippines and Indonesia, where they are illegally excavated from grave sites dating anywhere from 500 BC- 1,500 AD. These beads and bracelets are identical to the Palauan money, and probably came from the same origin. Whole groups of Palauans started travelling in search of these beads in the early 1990s. The antique dealers abroad quickly caught on, and the price of the beads and bracelets became higher than that of the gold beads and pottery they were originally digging the graves for.

Upset about the "new" beads, the Palau congress introduced a law 3 years ago to prohibit the importation of any beads into Palau. They could not get the law passed because there would be no legal way to prove which beads are originally Palauan, and which are imported, because there is no accurate inventory. The smaller types of beads only have a color and type name, and only large, unusual pieces, and bracelet parts have individual Palauan names. The Palauan way to test a bead is to rub it against your nose (the skin oil makes the pattern, color, and surface more clear), and to scratch a mirror or glass with it (the glass paste beads will leave a scratch).

Today you can see some women wearing bracelet pieces from abroad that are much larger than the largest ones which were known by name and famous in Palau in the past. A lot of the smaller "new" beads are already in circulation and used during customs unnoticed. Both the Philippines and Indonesia recently have passed laws to prohibit the grave digging, and a lot of the small antique shops have closed. So there is a good chance that the traditional Palauan money system, after a period of 15 years during which limited quantities of new beads and bracelets were brought into circulation, will go back to being a closed system.

Mengungau *orange*, Berrak *yellow*, Idek *red*, Merimer *glass inlay*, Cheldoech *glass*.

Merimer

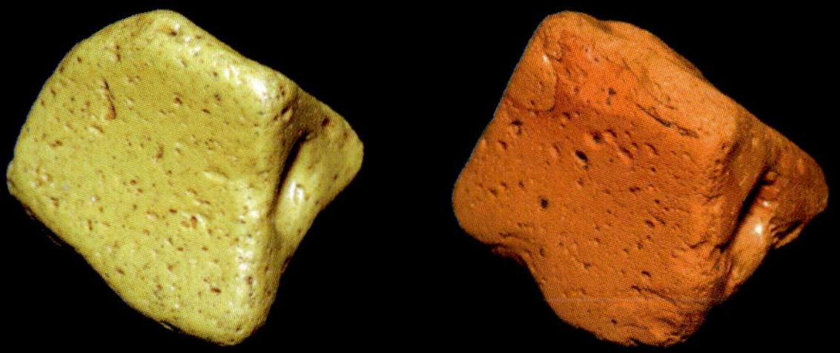

Dili bachel

Bachels are pieces of bracelets. Whole bracelets have not been seen in Palau since before 1900, but legends carved on the Bai meeting houses in the early 1900s show Palauans cutting bracelets into smaller parts to provide more Bachel pieces. Identical bracelets, probably from the same origin as the first Palauan pieces, are found in excavations around Asia dating to around 500 BC- 1,500 AD. Since most excavations are done illegally by farmers looking for gold and pottery to sell, there are not many records available. The bracelets are found in graves together with gold beads and Chinese pottery. They are made from a mixture of glass paste with mineral powder, the same material as the **Kldait** beads on the next pages. Their origin is unknown, since they were not manufactured in any of the countries where they are excavated, but it is likely they were manufactured in ancient bead centers in India or the Middle-east, and brought to Southeast Asia by traders. The gold and pottery never made it to Palau, but somehow the beads and bracelets did. What is interesting to note is that the **Merimer** type of bracelet, which is dark blue, green, or black with delicate colored glass inlaid swirls (see above sample) is the rarest type to find both on Palau and abroad, yet it is of lower traditional money value than the yellow or orange bracelet pieces of the same size on Palau.

cutting of the original bracelets and the drilling of two holes to be able to put a string through was done in the 1800s with primitive shell hand drills, and must have taken weeks, since the glass material of the bracelets is mixed with mineral, and very hard.

Interviewing antique dealers abroad, I found out that many of them had started cutting up the bracelets so they could charge more to their Palauan customers, but that it takes a lot of effort and a diamond drill bit to cut the bracelets. Shown below are bracelets of all the color types that are considered valuable on Palau. Only the plain translucent green or blue glass bracelets have not been considered valuable anymore since 1900 because the material is easily counterfeited.

Delomel busech

In early times the inhabitants of Ngeukl (Ngeremlengui) would drill an additional hole in the front of a Bachel money piece, and carve lines on the sides. A bird feather would be stuck into the hole as decoration. These pieces were called **Delomel busech** *(planting a feather)*, and there were only 4 or 5 in existence. The area of Neukl was destroyed during a war with neighboring areas, and only the name of the village and the money pieces remain. *Right, above: One of the original pieces from Ngeukl, worn by Rosania Ngiraikelau.*

Paramount Chief Reklai Rafael Ngirmang of Melekeok, the largest piece of money he is wearing here is the bachel called ***Kedam***.

For centuries, the two largest money pieces on Palau were the Bachels called Mchuchuu and Kedam, both yellow colored bracelet pieces. Kedam is thin and bright yellow, while Mchuchuu is thick and greenish yellow. Wars were fought over these pieces in the past between the two Paramount Chiefs of the highest clans of Palau, Chief Ibedul of the Idid Clan of Koror, and Chief Reklai of the Udes Clan of Melekeok. The only time the two clans inter-married was in the mid-1800s, when Chief Reklai of Melekeok married the highest ranking Idid woman, the Bilung of Koror. A peace treaty was finally signed between Melekeok and Koror in 1883, organized by the British warship Espiegle and visiting anthropologist Johann Kubary. Below are some of the first observations by foreigners on the Palauan money and its use:

June 1791

The Chief (Reklai) gave into the hand of a rubak a bead, which he carefully inclosed in his hands, and then moved slowly towards Abba Thulle (Chief Ibedul) with his body bent, as is usual in approaching the King; he said something in a low tone of voice, that seemed to meet the approbation of the assembly; he then appeared to be in the act of presenting this bead, and Abba Thulle on the point of receiving it, when he suddenly drew back his hands and asked, if so rich a present did not entitle the bearer to some reward; the King immediately gave him a smaller China bead; as soon as the rubak had received it, he, with great solemnity, resigned the rich present to the hands of Abba Thulle, who made a motion to retire.

Lieutenant Wedgeborough was naturally anxious to know what it was that had put an end to so much dissension and bloodshed; the King kindly indulged him in his request, at the same time enjoining him to be extremely careful, lest he should let it fall to the ground. This precious token of peace and amity was a bead or stone of a yellow colour, in length about two inches, in depth or thickness one inch or more, formed like a wedge with two holes at the base, for the purpose of putting a line to suspend it on.

From: A Supplement to the Account of the Pelew Islands. 1803- John Pearce Hockin
Based on the 1791 journals of the ships Panther and Endeavour.

October 1783

One of the Pelew (Koror) people on this occasion distinguished himself in a very extraordinary manner; he ran to the house while it was in flames, tore off a burning brand, and carrying it to another house, where many of the enemy had taken shelter, set it on fire, and it was quickly burned down; the man had the good fortune to return to his companions unhurt; the King (Chief Ibedul) publicly rewarded him for his courage, by immediately placing with his own hands a string of beads in his ear, and making him afterwards an inferior rubak on his return to Pelew (Koror).

From: An Account of the Pelew Islands, fifth edition. 1803- George Keate

Bilung Gloria Salii of Koror wearing the famous piece called **Mchuchuu,** traditionally the thickest yellow bachel found on Palau.

We lay at this place 3 days and nights, during the negotiations between both islands. The King (Chief Ibedul) was set upon the litter, and 8 men of Artingall (Melekeok) carried it in their hands off the pier, and then on their shoulders to the place of state. They enthroned him on a high seat made of wood, and covered with mats. The two first chiefs approached him, half bent, holding the jewels (beads) suspended by strings, and presented them to the King. One rubak only was removed from his office in Artingall. We re-embarked for Pelew (Koror), and took the broken rubak and 60 Artingall women with us.

From: Voyages and travels in the northern and southern hemispheres. 1817- Amasa Delano

1862

Thinking Aibukit (Ngaraard) would summon a warship to avenge the damages suffered, Ebadul (Chief Ibedul) now came in his own exalted person, and offered his former enemies peace and friendship. He did not return the stolen boats and objects, but a large piece of native money of the most valuable kind compensated for the loss completely, according to the custom of the country. The act of refusing to accept it would have constituted a declaration of war. The money was tempting to them also, for it was a piece of the berrak (yellow) type, which is most highly regarded, and there were only 3 or 4 of these large specimens in the whole country, and with it they would be able at any time to buy the life of any chief.

The question of where these pieces of money came from and who made them is of the greatest importance. Unfortunately, the investigation seems to be filled with so many difficulties that I doubt that this material can ever be correctly interpreted.

From: Die Palau Inseln. 1873- Karl Semper.

Kedam

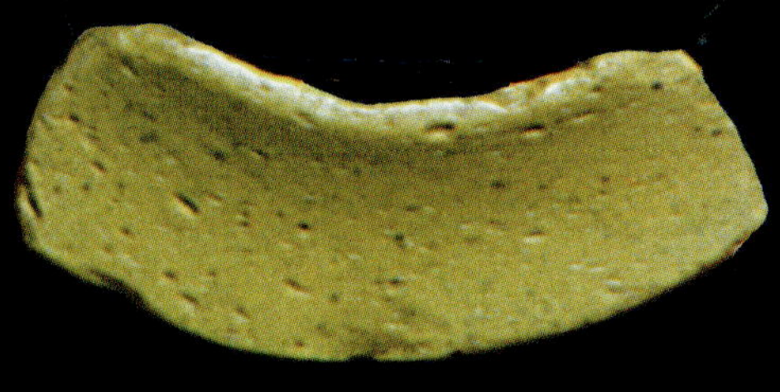

Mchuchuu

Chelbucheb

Also known as eye and mosaic beads, the technique and material to make these beads came from Egypt and the Middle East. Made in Eastern Java around 600 to 900 AD, these beads are only found as heirloom beads on Palau and Borneo, and some have been excavated in other areas of Indonesia and the Philippines. On Palau these beads were traditionally owned by the high ranking rubaks only. Sizes and pattern can vary, and the most valuable Cherecheroi type is called Meringel Chad when lined with yellow. Before the 1990s, there were only two unusual large-sized Chelbucheb beads on Palau, they were a famous set of earrings once owned by the Paramount Chief Ibedul, but taken by the Paramount Chief Reklai in the 1800s. Tearing it out of the ear of Reklai, Ibedul broke one of the two beads. The broken piece is still in the Idid Clan collection today. The other bead, which is of the same size but has a different design (see next page) was sold by Reklai Lomisang in the 1960s and is presently owned by the Etpison family. Recently, small and large types have been brought in from Indonesia, where every type of Palauan Chelbucheb has been looted from grave sites, found together with many other types of beads that were never known in Palau.

Kluk

Also called combed polychrome beads, these are found in many different cultures, and some types were manufactured in Indonesia in the past. On Palau, the Kluk is like the dollar value, everything is measured against it for value. One Kluk was worth 40 baskets of taro around 1880, and one Chelbucheb bead could be worth 5 Kluk beads. Daily life payments took place in beads up to the value of a Kluk, the more valuable money pieces "did not move", as the Palauans say, were carefully hidden, and were only used by Chiefs for important payments.

Kldait

Cheldoech

Strings of smaller beads made from this type of material are called Indo Pacific beads, widely distributed by traders from 200- 1,100 AD. Not much is known about these large sized Palauan beads. These are the bead types that are commonly used for traditional exchanges between all Palauan families. Made out of the same strange glass paste with mineral material as the Bachel bracelets, the Kldait beads are found in yellow, orange, red, and regular glass on Palau. The glass ones are still worn, but have lost their traditional value (left). What is interesting, is that the yellow and dark red beads, and Kluk and Chelbucheb beads are only excavated in Indonesia, while the orange beads and yellow and orange bracelets are only found in grave sites in the Philippines, where the common Indonesian bead colors are extremely rare.

Omebael

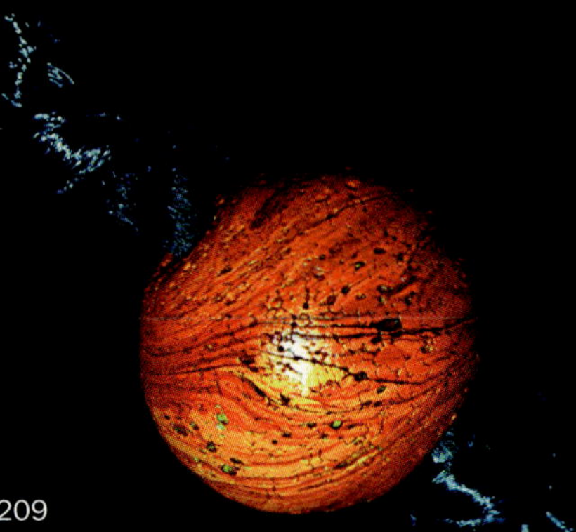

Traditionally and still today, when a girl is pregnant for the first time, the family of the baby's father will give her a large round bead to wear during her pregnancy. The bead is then called **Omebael** (to create), and is put to insure successful development of the baby. The bead is returned after the pregnancy. A smaller round or other shaped bead is often worn by young girls on a daily basis, and just called **Olbiungel**, this is the general name used for necklaces of any type.

Bleached *(yellow lining)*

Kldait

Mengungau *orange*, Berrak *yellow*, or Idek *red*

Bleob Kloranges Chesbad

Orabech Ibtaches Ertochet Elebangel Orak

Origin of the beads

Finding some clues as to the origin of the money-beads, and a time frame when they were brought to Palau could help shed some light on the origin of the Palauans themselves too. From recent research on beads *(Francis, 2002)* we know that the Chelbucheb type beads were manufactured on East Java between 600-900 AD, during which time there must have been some contact with Palau by traders, or settlers. The other bead types are more difficult to trace, and all we know for sure is that they were part of a bead making system that started in Egypt 4,000 years ago, and made its way to China, India and Indonesia around the time of Christ, and stopped around 1,100 AD. What is interesting is that the areas where those Palauan moneybead types are found, namely Indonesia, Philippines, Malaysia and possibly Madagascar, are also the areas linguistically closest related to Palau.

The largest rare excavated bead types from Java, Indonesia.

The island of Yap also offers an interesting angle. Kubary reported in 1885 that the Palauan moneybeads may have been brought in by the Yapese who founded Keklau village in Ngaraard State, or by a rich Yapese who married into the Idid Clan of Koror, and that the Yapese used to use this money themselves before switching to their current stone and shell money. Yapese had been in regular contact with Palau since at least the early 1700s, and the beads were reportedly found in the ground on Yap sometimes by accident during farming. It is not clear if they were found loose or in grave sites. Palauans themselves only have mythical legends about the origin of the money, referring to a sunken or floating island, or a bird as the origin. In 1882 Kubary was told by the National Museum in Tokyo, after showing them some Palauan beads, that similar beads had also been excavated in Japan, but were never manufactured there. The Japanese believed they were brought by the first Buddhist pilgrims from Asia.

The Dayak tribe on Borneo considers certain beads, some identical to the Palauan moneybeads, valuable also. They are sewn onto baby baskets to protect the baby against evil spirits, and are worn on strings during special occasions. The few old Chinese records that survive show beads being brought to Borneo and the Philippines by Chinese traders between 1127- 1650 AD, and Indo Pacific beads have been found in Chinese and Korean tombs dating from 300 BC- 600 AD.

The early beadmaking centers and grave sites on Java, Indonesia, show a connections with South Indian bead centers and burial practices. The contemporary Indonesian beadmakers still know the ancient techniques to imitate the Chelbucheb, Kluk, and plain glass type beads perfectly. These beads are sold to Japanese and European collectors looking for the ancient excavated beads. They do not know the material or how to make the Kldait type beads though, so these were probably not manufactured on Indonesia originally, even though they are excavated there. The Indonesian culture and language has many similarities to Palau, and could be where some of the settlers of Palau and Yap came from.

On the Philippines, the orange Kldait beads, and bracelets of all colors are excavated, but were never manufactured there. Chelbucheb, Kluk, and yellow beads are occasionally excavated, but rare. Only two bracelets of the Merimer type were ever excavated in the Philippines, and are owned by a private collector in the Philippines. Thin flat translucent bracelets with the same color glass inlay designs have been found also.

Farmers on the islands of Samar and Cebu in the Philippines originally looted grave sites to sell the gold beads and pottery to antique dealers. The large orange beads found were usually strung into a necklace and sold for around $20 per string of 50 pieces in stores in 1990. The bracelets sold for $30 each, while the Chelbucheb and Kluk beads were $50, since they are rare there. I recently checked with the antique dealers in Manila, and was amazed to find out that the orange beads now sell for $50 per piece, and the bracelets are up to $1,000 each because of the Palauan demand. The Chelbucheb and Kluk beads have not changed in value, since Palauans discovered them in Bali, where up to last year they were a lot more common and cheaper than in the Philippines. Now most antique shops on Bali have closed down, and the grave digging on Java is strictly prohibited by the government.

While working on western Cebu in the early 1980s, I once witnessed a grave dig, but I was not aware of Palau or their beads at the time. Grave digging was done there by local kids and farmers, who wanted to earn some extra cash. They would poke sticks into the ground until they could feel the location of a grave, usually by hitting a piece of pottery. The graves were surprisingly shallow, less than 3 feet deep. The one I saw contained 3 small pieces of white and blue pottery, one of them cracked by the roots of a nearby tree. The face was covered by a large green ceramic plate, and they found only a handful of orange colored and finely made small gold beads. The boys told me that this was an average find, but on rare occasions they would stumble on a grave with larger gold pieces such as a face mask or eye covers, or gold bracelets and whole necklaces. These were sold to antique dealers in Cebu city, who in turn sold it to Manila retailers.

So even though Palauans have been bringing in beads and bracelets from abroad over the last 15 years into their previously closed money system, the quantity was small due to the limited availability, and the origin of the beads, still not quite clear, was probably the same as that of the Palauan beads. Now, most of those imported beads, especially the small ones identical to the original Palauan money, are already in circulation on Palau, and have been used repeatedly during family exchanges in recent years.

Modern Palau

The three generations of Palauans alive today were born and raised in totally different ways. The older generation was born during the Japanese time (1914-1944). A lot of them learned to speak fluent Japanese, and were raised in poverty with forced labor and little education, and the traditional way of life of their parents was strongly discouraged. After the war they had a difficult time adjusting to speaking English, and the new, strange American ways and canned food. Their children were raised as Christians, speaking English in the local missionary schools, some going on to high schools and colleges on Guam, Hawaii, and in the States. There was a huge generation gap in education, thinking, and way of life between the two, and a lot of the children did not return to live on Palau, but found jobs and made their lives abroad. For both generations, the change of diet to rice and fried foods strongly increased diabetes, cancer, and cardiovascular diseases. Almost 80% of the population had Hepatitis B before they started giving mandatory shots at birth in the early 1980s.

The computer-age teenagers on Palau today have more than 30 television channels, and are used to air-conditioning and American junk food. They are being taught about American Indians in school rather than about their own history, and most of the traditional craftsmanship and knowledge is gone. Palau today is a strange mix of Japanese, American, Christian and traditional Palauan influences.

The good thing is that the young Palauans seem to have a renewed interest and pride in their own history, and the need for this is starting to be recognized by the present leaders. The old German records are starting to be used to re-introduce some of the lost information, and a new Palauan history book was published and is being used by the public high schools. The Japanese and the missionaries taught the previous generations that they had to change, adapt to foreign ways and forget their old traditions, which were no good according to them. Now this needs to be reversed to teach the young people that everything foreign is not necessarily better for them or their country.

Even now, some older Palauans are still in denial about their past, and don't want to know that their grandfathers used to run around naked and cut off heads, and that their grandmothers used to use prostitution as an income for their villages. They are ashamed of these things now, and try to live as good Christians. In private however, a lot of them still believe in some of the old superstitions and beliefs of their ancestors, and they still make offerings to the Spirits and consult with magicians.

They do not realize that they can be proud of their ancestors with their simple tools, who were able to achieve amazing pieces of architecture and art, evidenced by the Bai meeting houses and canoes of the past, and the inlaid wooden pots they used to make, which no one can do the same way anymore today. The young Palauans have to learn to be proud of their past, and to recognize that the culture of the 1800s is what made them what they are today, not just the foreign influences. Hopefully they will try to keep some of the unique parts of their culture alive before these disappear forever.

A lot of Palauan families today have mixed Japanese, Chinese, and American blood, but since the Palauans have always welcomed foreigners to marry into their clans since the first European ships came to the area, it is hard to say if there is such a thing as a full-blood Palauan. The first thing foreigners noticed about the Palauans in the 1700s was the variety in skin color and hair texture, even in those early times, so Palau was probably already a melting pot of cultures before any of the major foreign influences of the 1900s took place.

The hardest thing this new generation will face is how to become truly independent and run their small country, without letting foreign nations take over, and without letting foreign industries come in and take advantage of the natural resources, or damage the environment, which are Palau's greatest assets for the future. The politicians of today are selling the country too cheap, and Palauans have to make sure there will be something left for their kids to be proud of.

TEAM PLAYER

Politics

Traditionally, each village on Palau had its chiefs, and a council of chiefs met regularly for important decisions on wars and rules to keep harmony and peace in the community. Customary punishments for individuals, clans, or villages were in the form of food or moneybead payments, and even a human head at times.

During the German and Japanese times (1899-1944) foreign laws started to take over, and local chiefs didn't have their traditional power.

Palau became a Strategic Trust administered by the United States in 1947, under the Trust Territory of the Pacific Islands. Courts and a Palau Congress were established. The Secretary of the Navy administered Palau until 1951, when it was transferred to the US Department of the Interior until October 1994, when Palau became independent. In 1981 Palau adopted its constitution, and Haruo I. Remeliik became the first elected President. In 1982 Palau signed a Compact of Free Association with the US, but it was not approved until 1993, after local unrest, seven votes and finally an amendment to Palau's

Although local warfare does not take place anymore, and no more heads are cut off, many people say the Palauans now fight it out in the political arena instead, and Palau has been called the most over-governed place on earth.

It is amazing how complicated they have made it to run such a small country with only 19,000-some people. There are 16 tiny states, each with their own legislatures, budgets, laws, Governors, and traditional village chiefs. That's how the Americans set it up after WWII. Palau also has a US-style democratic National Government and congress, as well an advisory council of traditional chiefs. The President and Vice-president run separately, although they are trying to change that at the moment, since it does not work very well. The congress has a House and a Senate, and elections are every four years.

The competition between clans, individuals, and states for money and power is as strong as it ever was in the past, and the congressmen are usually too busy fighting amongst themselves to get much done for the country's long-term benefit.

The funny thing about Palau's elections and politicians is that most of them do not have any platform, and even if they do, the majority of the people vote based on who they are related to, rather than on specific issues. This is slowly changing with the younger generation today.

Palau is facing a financial crisis, since the Government spending is higher than the revenue coming in from the local industries such as tourism, trade, and commercial fishing. With the US Compact funding reducing over time, the only way Palau is going to survive the next ten years is by making a smaller, more efficient government structure, support and allow their private tourism sector to grow, and by getting more foreign assistance and grants.

Since Palau has now a full vote in the United Nations, foreign powers have been eager to make deals with this tiny new nation. Aside from the US Compact funding for the new Babeldaob road and the Melekeok Capital building, recent foreign grants to Palau have paid for the new bridge (Japan), docks for all the local states (Japan), additional costs of the new Capital building (Taiwan), and for a new Cultural Center and National Museum to open this year (Taiwan).

Problems in paradise

For visiting tourists, Palau seems like a tiny paradise, with the exotic beauty of the Rock Islands, coral reefs teeming with marine life, and only a handful of people. But like all developing nations, Palau has its own unique set of problems, and is making a lot of the same mistakes other countries have.

Having extended family clans to fall back on, no one on Palau goes hungry, and the ambition to work hard is lacking in all but a few. A lot of Palauans seem to think that if you work harder and earn more, your relatives will just hit you up for more money during customs, and you will end up spending it all anyway...The traditional complicated exchange system of food and money between relatives, which was designed to share responsibilities and help each other out, has now become a burden for young people, who do not get a chance to save money for their own future and kids because of family obligations, and have to take time off from work to assist with family customs. That is one of the reasons a lot of young Palauans choose to live and work abroad.

Foreign labor is brought in by local companies from the Philippines, Bangladesh, Taiwan and China, including gardeners, maids, waitresses, sales persons, and construction workers. American lawyers flock to Palau since they have discovered that Palau's laws are based on the US laws, so it's easy for them to pass the bar, and there are enough local land title disputes to keep them busy for years. The land title disputes date back to the Japanese time, when a lot of land was confiscated by the Japanese. After WWII it was supposed to be returned to the original owners, but they didn't get around to it till the 1980s, by which time most of the original owners were dead. Cousins, kids and grandchildren now fight over pieces of land, government surveys are notoriously slow, and matters are complicated by the fact that a lot of the land is clan land, owned by a whole family, which means you are dealing with dozens of people claiming the same piece of land. The tradition of adopting kids within families, and one's ranking within a clan can complicate things even further.

Tour and retail businesses by law should be 100% owned by Palauans. This was designed to encourage small Palauan businesses. In reality, all these businesses are being taken over by foreigners, who are either married to a Palauan and put the business in her or his name, or pay a Palauan Partner a few hundred dollars a month to own the business on paper, while the foreigner runs it. Some don't pay taxes, and this "front" business problem has been going on for years now, hurting the legitimate local businesses and government revenue.

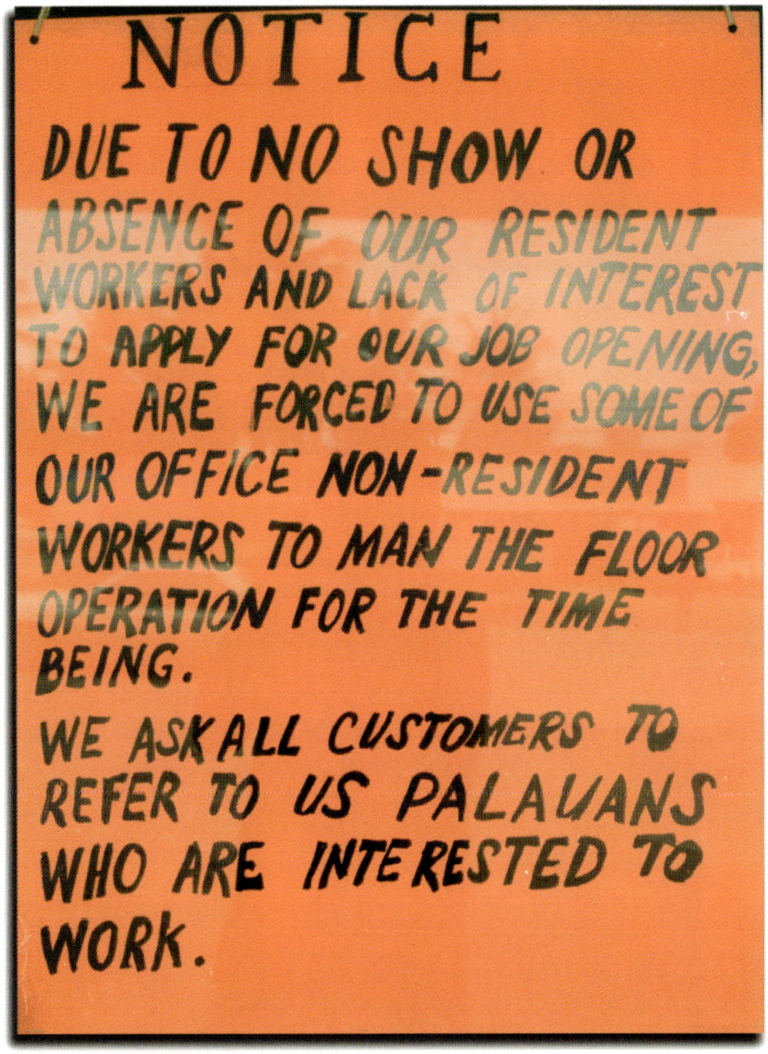

Although the speed limit is 35 miles per hour, and you are not supposed to overtake any car that is not standing still, traffic is becoming a real problem on the island of Koror. A good public transportation system mandatory for government workers, and stricter car laws on the import and quality of the cars would solve a lot of the problems. Third-hand unsafe cars are being imported from Asia and sold cheap, and the abandoned car bodies are starting to pile up and cause a trash problem for the country. Most families have two cars to drive up and down the one mile road, and biking is unsafe because of lack of bike lanes, aggressive feral dogs, and drunk drivers. Traffic lights were introduced, ignored, and abandoned after only a few years, and police officers direct traffic now during the "rush" hours of dropping kids to school at 7.30 AM, lunch hour 11.30-12.30, and the 4.30 PM rush to get home, when traffic is bumper-to-bumper on the main road.

After government workers get their two-week paycheck, there is usually a rush on the bars at night, and drunk driving, alcoholism, and domestic violence is still a big problem. There was a midnight curfew established in 1985 on Koror after President Remeliik was murdered, when all the bars had to be closed by midnight, and police patrolled the roads. The curfew was only recently cancelled after Continental Airlines started their late-night flight schedule.

Above: Sign in the window of a locally owned retail store.
Below: To see a car in the water during government "Payday Weekend" because of drunk driving is not unusual.

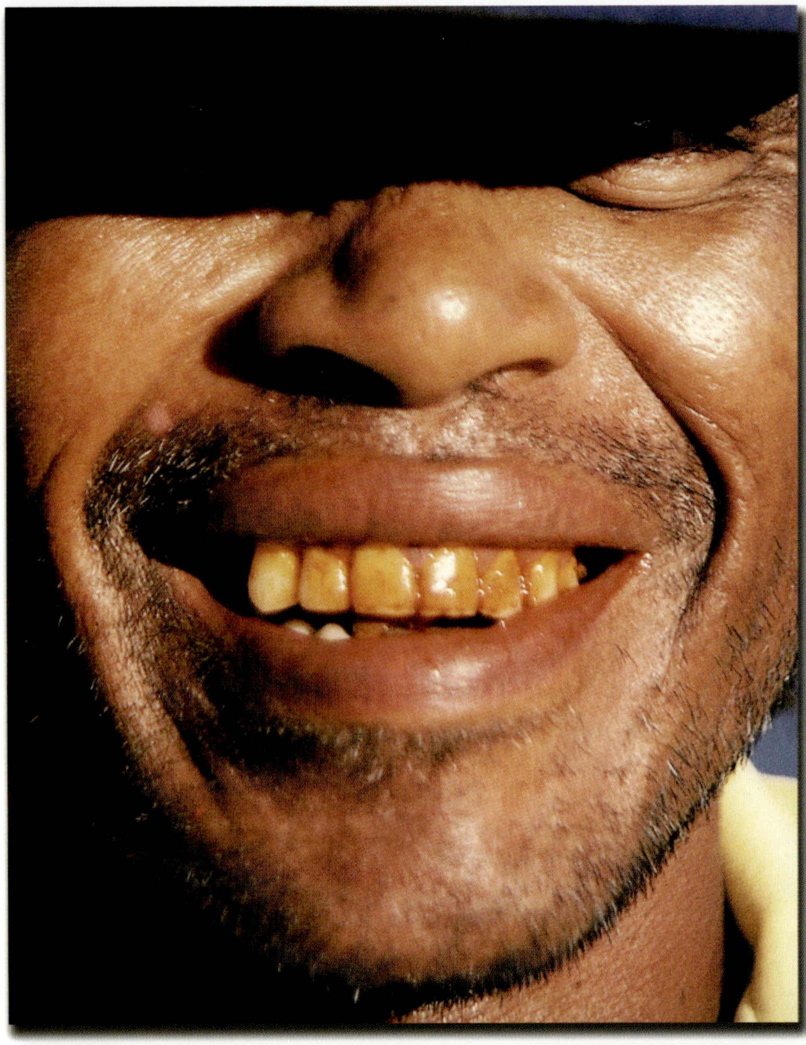

Betel nut

Chewing betel nut is popular throughout Micronesia, and probably 90% of the Palauans do it. The chew consists of the meat of the nut of the Areca palm *(Areca catechu)* **Buuch**, the leaf of the tree-climbing betel pepper vine *(Piper betle)* **Kebui**, and lime made from burned coral and shells. Aside from being a personal habit and addictive, it is a part of social life. This modest stimulant produces a high when chewed continuously, and reduces hunger and fatigue, but can also cause cancer and tooth decay. Evidence of betel nut chewing can be seen all over town, since they often spit the blood-red saliva on the sidewalks and streets, something building owners do not appreciate, since removing the stain is difficult. Palauans were already chewing when the first Europeans arrived in the 1700s, and carried their betel nut around in baskets, with bamboo limeholders.

Left: The teeth...
Below: The ingredients...
Next page: Hanging out and chewing.

NSG-2

Fishing

Fishing is the national hobby for Palauan men. Bottom-fishing, spearfishing, trolling, or casting, that's what most boats are going out for every weekend, and often families spend the night at the Rock Island beaches. Pairs of snorkelers go out at night to spearfish, with cut drums floating behind them on a rope to put the speared fish in before the sharks get them. The women usually stick to bottomfishing, go out on the reefs at low tide to get seacucumbers and sea urchins, and dig up small clams from the water's edge. They put coconut traps around the island during the day, and check for Coconut crabs at night.

Palauans usually eat all the fish they catch, or fish for specific customs, when the catch is shared by the whole family. Commercial spearfishing using scuba and export of certain species, as well as increased tourism and local restaurant demand for seafood have become a recent problem. This has resulted in several new laws.

Deepwater snapper *(Aphareus rutilans)* **Metngui**

Bonefish *(Alblua vulpes)* **Suld**

Malabar snapper **Sebus** *(Lutjanus malabaricus)*

Deepwater snapper (*Etelis carbunculus*) **Sebus**

Mahi-mahi *(Coryphaena hippurus)* **Ersuuch**

Spiny lobster *(Panulirus longipes)* **Melech**

Mangrove crab *(Scylla serrata)* **Emang**

Coconut crab *(Birgus latro)* **Ketat**

Ornate spiny lobster *(Panulirus ornatus)* **Melech**

406 lbs. Black marlin *(Makaira indica)* **Meluis**

Sailfish *(Istiophorus platypterus)* **Tekrar**

Great barracuda *(Sphyraena barracuda)* **Ai**

Giant trevally *(Caranx ignobilis)* **Cherobk**

211 and 208 lbs. Blue marlins *(Makaira mazara)* **Meluis**

Being able to swim with reef sharks is one of the main attractions for scuba divers on the outer reefs of Palau. There has never been an attack by reef sharks on tourists, since there is plenty of fish, and it is usually the divers who are chasing the sharks with cameras.

The opening of the Taiwan tourist market to Palau, and the 100-some long-line fishing vessels operating on Palau started causing problems for the sharks. The Chinese fishing vessels started catching sharks aside from the tuna they are supposed to target, to supply the high demand for shark fins by the local restaurants, as well as for export. Their vessels were often seen close to the southern reefs, while they are supposed to stay out in open ocean, and they use steel leaders on their hooks specifically for catching sharks.

Palauans are not fond of sharks, since they usually get chased by sharks when spear fishing at night, and sharks often take their fishing lures when they are bottom fishing, so it took several years to get a law passed to protect the sharks. A Micronesian Shark Foundation was formed, and is trying to start doing more research on sharks. The Palau Shark Sanctuary Fund was started to make people aware of the shark finning problem, and to lobby the public and congress for laws to protect sharks in Palauan waters. President Remengesau set an example in May 2003 by burning a confiscated catch of shark fins in public, and the law was finally passed in September 2003, making it illegal for foreign vessels to use steel leaders and to catch or in any way mutilate sharks and marine mammals.

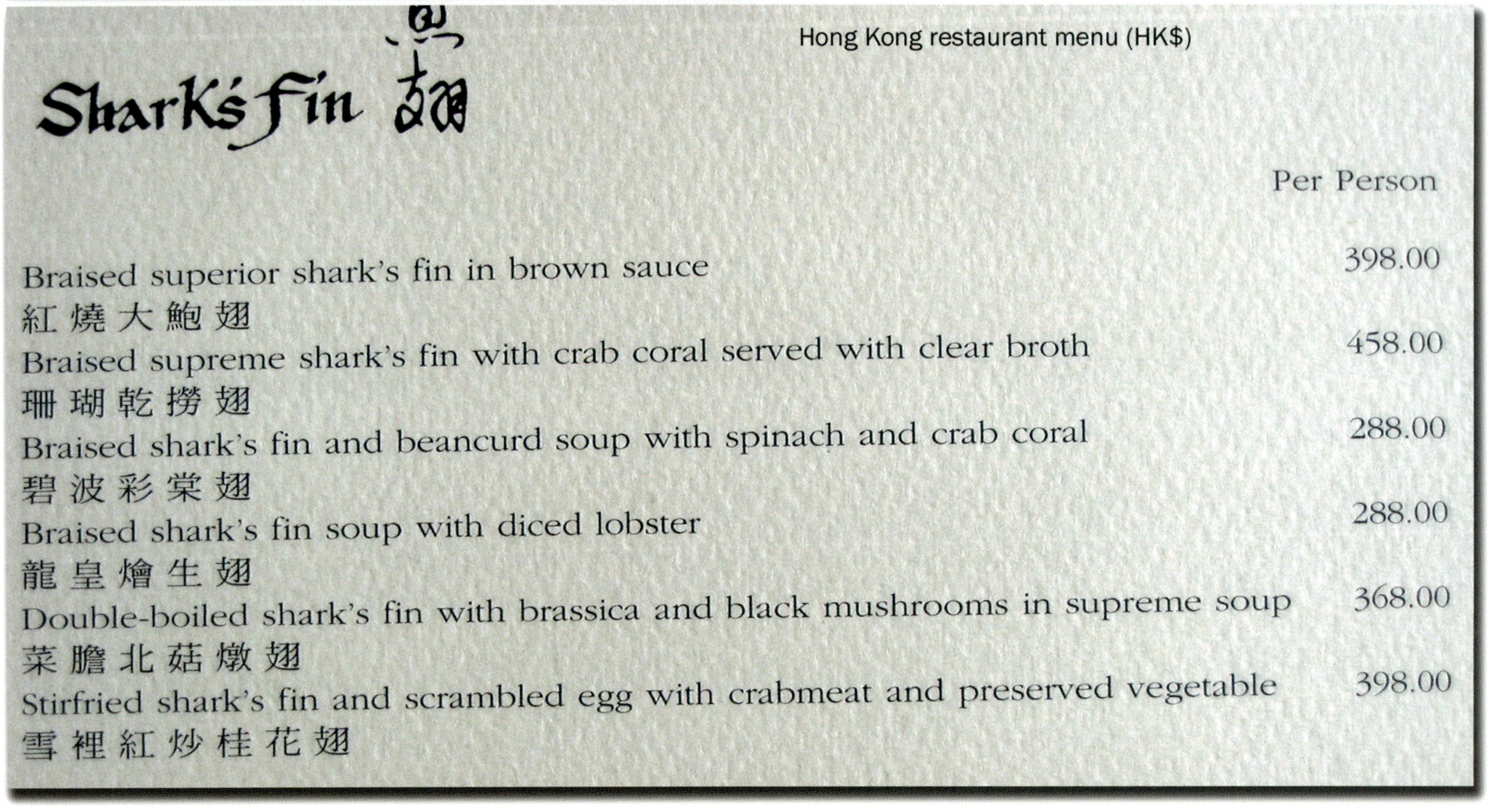

Hong Kong restaurant menu (HK$)

Shark's Fin 魚翅

	Per Person
Braised superior shark's fin in brown sauce 紅燒大鮑翅	398.00
Braised supreme shark's fin with crab coral served with clear broth 珊瑚乾撈翅	458.00
Braised shark's fin and beancurd soup with spinach and crab coral 碧波彩棠翅	288.00
Braised shark's fin soup with diced lobster 龍皇燴生翅	288.00
Double-boiled shark's fin with brassica and black mushrooms in supreme soup 菜膽北菇燉翅	368.00
Stirfried shark's fin and scrambled egg with crabmeat and preserved vegetable 雪裡紅炒桂花翅	398.00

Palauans normally do not fish for sharks, but they are caught accidentally when bottom fishing or trolling. The Grey reef shark below has a lure stuck in his mouth, but survived. The Pelagic thresher shark above was caught by Palauans who were bottom fishing for other fish at night in 400 feet of water. Traditionally, young Palauan men of certain districts had to go outside the reef to catch a shark with a coconut rattle and noose once, to prove their capability as a warrior and fisherman, but this has not been done for over a hundred years.

Hunting

Regular guns are not allowed by law on Palau, but hunting with airguns is very popular. Palauans hunt for fruitbats, pigeons and other birds. Spearfishing is also popular, and favorites include lobsters, turtles, and any type of edible fish. During funerals and other traditional customs, male relatives are expected to go out fishing to provide food for the occasion. Pigs are killed, and turtles are caught or speared, and tied up alive for days waiting for their turn to be cooked.

Unfortunately, a lot of the species the Palauans are fond of eating and hunting are now threatened, and need to be protected or they will not be around for much longer. This has been a difficult message to get through to the Palauans, who have enjoyed their fishing and hunting for centuries, and do not really want to look at the long-term effects.

Tourism, Palau's main industry, has helped to make the Palauan people and government look at these issues from a different angle, and the younger generation now starts to realize that they need to protect their nature and wildlife, which is the main attraction for tourism, and which in turn may be the only thing to keep their economy going once the US Compact funds run out.

Left, middle: A large speared green turtle lies tied up alive in the hot sun, waiting to be killed for a funeral custom.

Left, bottom: A local trap made up of twigs using coconut as bait to trap ground-dwelling birds.

Next page, top: Some locals still hunt the Saltwater crocodiles. They spot the crocs along the mangrove edges at night with flashlights, the light makes the croc's eyes visible. They get into the water and spear the croc in the tail, and throw a noose around the mouth before dragging the animal to the boat. The largest crocodile I have seen on Palau was this 16 feet long giant, which was caught about 15 years ago, when they came to borrow our forklift to lift it out of the water.

TARIMEL ESBEI
Forever In Our Hearts

DIVI

Threatened species

The species of crabs shown here are the most popular ones used for food, both by locals and in restaurants. The Coconut crab is slow and easy to catch by using tied coconuts as bait. The land crabs come to the beaches once a month to lay eggs, when Palauans can pick up dozens at a time. The mangrove crab is in high demand, and sold for as much as $45.- each in some of the restaurants, they are caught in traps under water near mangrove areas. The Pigeons are shot with airguns, and are considered a delicacy by Palauans. The Megapode itself is not hunted or eaten, but the eggs, which are dug out from the huge nest mounds, are also a local delicacy.

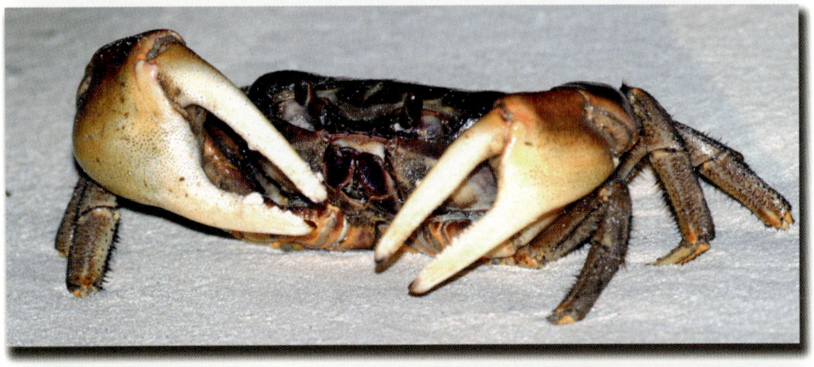

Left, from top to bottom:

Coconut crab *(Birgus latro)* **Ketat**

Blue land crab
(Cardisoma hirtipes) **Rekung el beab**

Mangrove crab *(Scylla serrata)* **Emang**

Micronesian Imperial Pigeon
(Ducula oceanica) **Belochel**

Nicobar Pigeon
(Caloenas nicobarica pelewensis) **Laib**

Palau Megapode
(Megapodius laperouse senex) **Bekai**

The Napoleon wrasse is a beautiful fish, and can grow to 6 feet long. It is a friendly and curious fish, and can be hand-fed by divers around Ngemelis. Unfortunately, it is also a favorite Palauan food fish, and locals go out and spear these fish regularly at night.

The Bumphead parrotfish swim in schools, and on rare occasions you can see a huge school passing by, but they are now shy. At night, they sleep on the reef and make easy targets for spearfishermen on scuba, who can get a whole boatload full in one night. It is now illegal to use scuba for spearfishing, and illegal to spear around the dive areas.

The Dugong is endangered in every Pacific island where it is found (Indonesia, Philippines, New Guinea, Palau). In Palau it was hunted for its meat as well as for the bones, some of which were used as valuable bracelets in the past. There are very few left on Palau, and are now an endangered species protected by law.

The Hawksbill turtle uses the Rock Island beaches as their main nesting site. Palauans dig up the eggs and also hunt the adult turtles for their meat and shell. Traditional money plates are made from the shell, which made it difficult to protect these turtles by law. There is a season and minimum size, but this is still not strictly enforced.

Napoleon wrasse *(Cheilinus undulatus)* **Maml**

Bumphead parrotfish *(Bolbometopon muricatum)* **Kemedukl**

Dugong *(Dugong dugon)* **Mesekiu**

Hawksbill turtle *(Eretmochelys imbricata)* **Ngasech**

Giant clam *(Tridacna gigas)* **Otkang**

Conservation

If Palauans are to protect their natural resources and unique native species of wildlife, there is a lot of work to be done. Since WWII, the Palauans have hunted and fished everything, without the traditional restrictions and respect for the animals that their fore-fathers had. Before 1900, pigeons, dugong, pig and turtle were only eaten on special occasions or village feasts, and only certain chiefs and villages were allowed to hunt for those species, and would exchange them with other villages for items they did not make themselves, like clay pots or oil. Since WWII, traditional restrictions were abandoned, and the increased use of speedboats, scuba and guns made sure the local wildlife didn't have a fighting chance. Export of certain species brought them close to extinction before any laws were passed.

Local kids need to be educated, so they will not make the same mistakes as their parents' generation did. Since they do not learn it at home, better elementary school programs and special fieldtrips are needed.

The problems need to be solved quickly to save some of the already endangered species. There is no problem with Palauans going out hunting and fishing as they used to do, if they follow the allowed seasons and sizes, don't spear-fish on scuba, and protect a few species like the Hawsbill turtle and the Dugong, which are already endangered. Commercial fishing should be limited and charged more, and export and consumption of illegal and threatened species in local restaurants needs to be more strictly enforced. The increase in tourists in recent years caused more demand for local seafood, which has already put a strain on the local reefs, and several restaurants have been fined for selling undersized lobsters and crabs. These species should be protected or imported for restaurant use, until a sustainable aquaculture industry is established locally.

For more information on local laws and conservation efforts, you can contact:
Palau Conservation Society- P.O.Box 1811, Koror, Palau PW 96940 Email: pcs@palaunet.com

Above: The nest and two chicks of the Micronesian honeyeater **Chesisebangiau.**
Next page: My son Iked holding an adult endemic Fruitbat **Olik**, this bat was raised in captivity after the mother was shot.

Tourism

Tourism seemed to be the answer for Palau's economy in the 1980s, until the focus shifted from high-end European,Japanese and American scuba divers to cheap Taiwanese group packages. September 11, SARS, and the bombing of a tourist bar on Bali reduced tourism to the whole region including Palau for almost two years, but it is now recovering. Being politically stable, using US dollars and English language, and not having many strange tropical diseases all give Palau an edge over many other Pacific and Asian destinations. The coral reefs, sharks, and fish life of Palau are famous among scuba divers and snorkelers, and the beautiful Rock Islands keep kayakers, photographers and nature lovers coming back. Although not very well known yet, Palau has an unusually high repeat tourists percentage compared with other destinations.

The problem Palau faces is that it is far and expensive to get to, and environmentally it cannot support the Asian mass-tourism it is starting to receive, without damaging the unique nature and wildlife tourists come to admire. The answer would be high-end eco-tourism, but the lack of high-end hotel facilities, the increase of Taiwanese tourists, and the limited flight connections to the States and Europe have resulted in the opposite. The last El Nino phenomenon hit Palau and most of the Pacific in 1997-98, causing droughts and high water temperatures. Shallow coral reef colonies died because of the water temperature changes (coral bleaching), and even most of the jellyfish in the famous Jellyfish Lake died. The deeper reefs and Jellyfish Lake have recovered well, but shallow corals as seen below in a pre-El Nino photograph at low tide, do not exist anymore, most of these exposed reefs are now covered with algae and will take many more years to grow back. Luckily Palau was not hit as hard as some of the other Pacific Islands.

Corruption in the National and state governments, ever-changing laws and regulations, and land title disputes scare off legitimate investors, and make high-end hotel developments a real challenge, because no Palauan can afford to do this type of project without foreign investment. The Palau Pacific Resort, a 160 room hotel on Ngerkebesang, Koror, opened in December, 1984 and is now, 20 years later, still the only beach resort on Palau (see page 43). Smaller and cheaper basic hotels, motels and apartments can be found on Koror and some of the other islands.

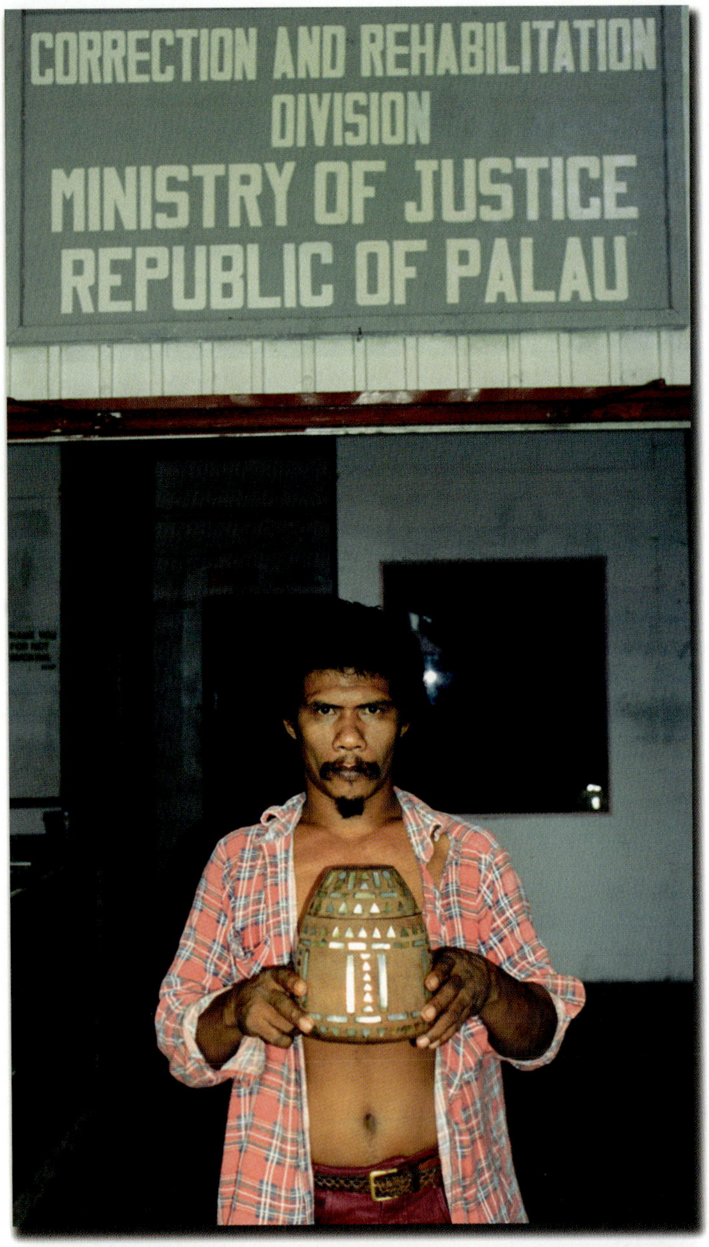

Storyboards

Palau is known for its carved Storyboards, which show some of the local traditional legends. In the past these were carved and painted on the traditional meeting houses, and were a way to record the local history. The storyboard souvenir started during the Japanese occupation in the 1920s, and is today still popular. The best carvers used to be in the local jail. Some of the prisoners in the 1980s didn't want to leave, and spent most of their life carving. Today, there are several stores in town selling good quality storyboards also, and you have a lot of choice.

When buying a storyboard, make sure what type of wood it is carved from. If there are small round holes anywhere on the wood, it may have termites and you should not buy it. Some of the local hardwoods are beautiful, but will crack badly when brought to cold or dry climates. The safest bet is mahogany wood, which does well anywhere. Most of the storyboards are not varnished, but rubbed with shoe polish, so you may have to refinish them periodically at home when they get stained or fade. The store or person you buy your storyboard from should supply you also with a story explaining the carved legend.

There are about 20 of the most popular legends that are being used, like the story about two lovers meeting on Ngemelis beach, and discovering the egg-laying cycle of the turtles. Another one is about the sunken village of Ngibtal, with fish washing out of the branches of a Breadfruit tree. A little book with most of the local legends is for sale on Koror.

Sometimes if you are lucky you can find miniature canoe models, Tobi Island monkeymen, or other carvings which are not available for sale on a regular basis. Pieces with shell inlay are now expensive and difficult to get.

Other souvenirs from Palau include simple woven baskets made from pandanus leaves, imitation Palauan moneybead necklaces, and shell jewelry.

Be aware that even though some local stores still sell turtleshell jewelry, and it is legal on Palau, it is made from the worldwide endangered Hawksbill turtle, and it is illegal to import this into most countries. You would also help the local Hawksbill turtles survive by not buying any of this jewelry !

Next page: Some examples of storyboards, a Tobi Island monkeyman, and local carvers at work.

MS-152

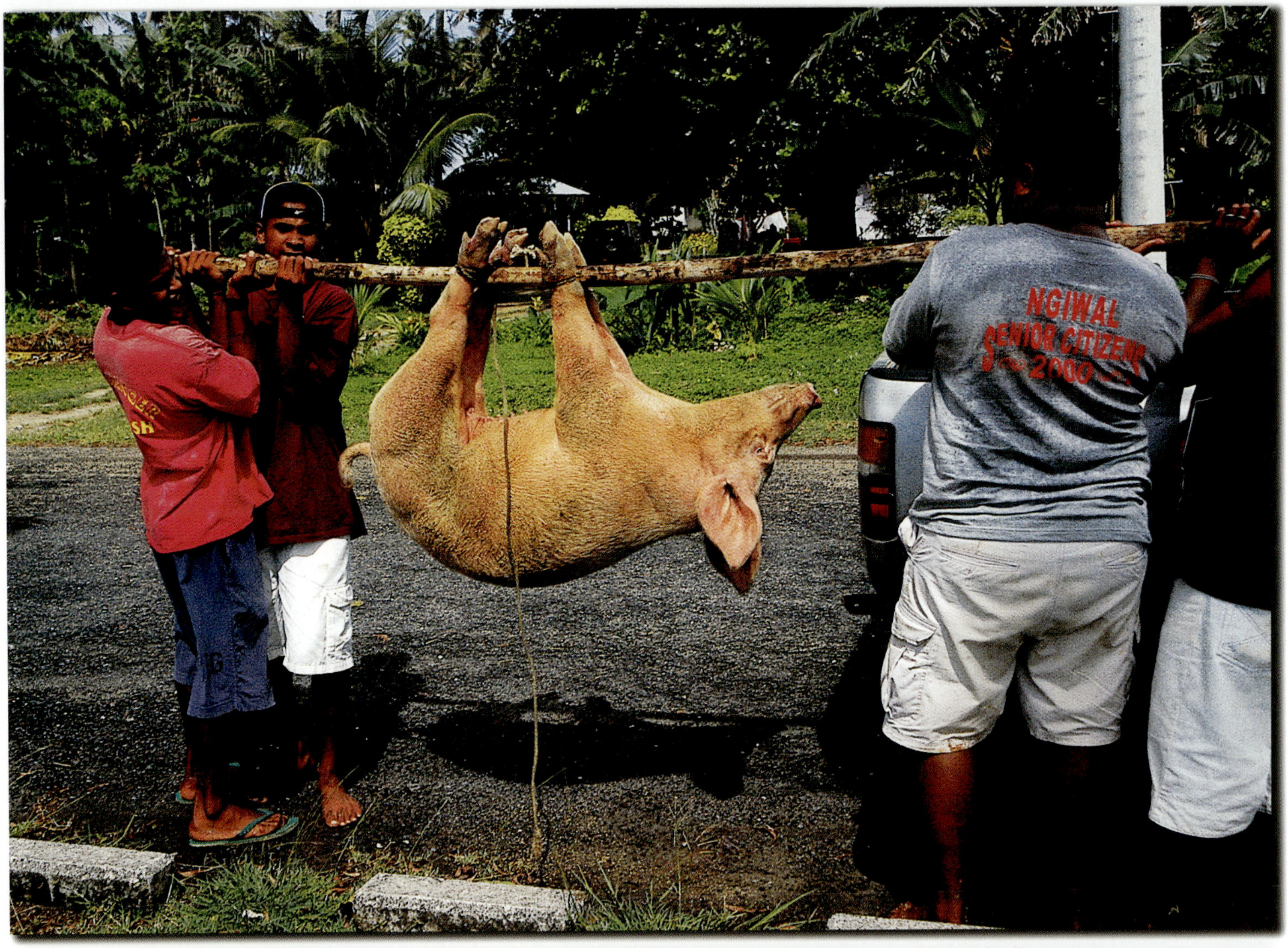
NGIWAL
SENIOR CITIZEN
2000

POLICE
POLICE
POLICE
POLICE

FROOT LOOPS
KIKKOMAN
Soy Sauce
BLACK FLAG
Coffee Creamer

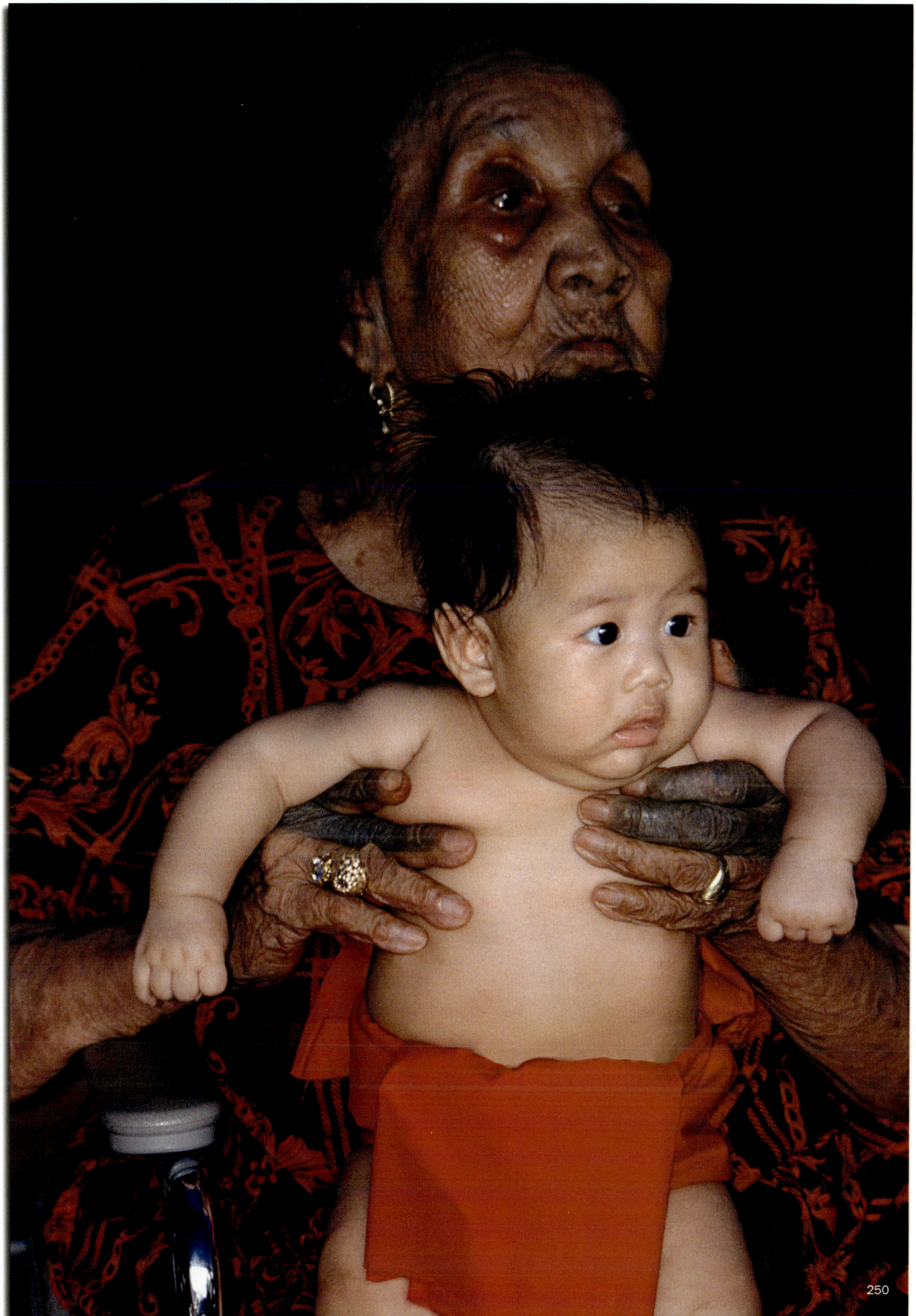

Some of the photographers...

Kevin Davidson

Kevin came to Palau in 1991 on the Palau Aggressor live-aboard, before working with NECO Marine dive center. He is an equally good dive instructor, captain, fisherman, and cook, but his real talent is as a video- and photographer. His beautiful images have been published in many magazines and books. He is known as an underwater photographer, but few people realize he does just as well above water with island scenery shots. Kevin now runs his own KD photo shop with his wife Tomoko, and is hoping to expand and start selling his images on-line soon. Kevin co-authored the dive guide "Palau-dive sites, history & culture" in 2001.

William E. Perryclear

Bill came to Palau in 1990, the Pacific islands always having been his dream. He did a photo project for the Belau National Museum for one year and a half, and set up his photo business, part-time at first. He also worked with the Senior Citizen Center for several years, starting up their gift shop, as well as a canoe building project. Inspired by local Maryland newspaper photographer A.Aubrey Bodine, Bill has always wanted to capture images that are both fundamental and thought provoking. Bill has published several books and many postcards on Palau, and did the photography for the book "A personal Tour of Palau" by Ann H. Kitalong.

Patrick L. Colin

Pat is a marine biologist, writer, photographer, inventor, pilot, and President of the Coral Reef Research Foundation on Palau. Since receiving his Ph.D. from the University of Miami in 1973, he has lived in Puerto Rico, the Marshall Islands, Papua New Guinea, and the Federated States of Micronesia, before setting up the Research Foundation on Palau in 1993 with his wife Lori. Based on Palau, they travel all over the world collecting marine invertebrates for medical research. Pat's hobby of flying his home-built plane while taking photographs at the same time has provided a wealth of aerial photographs on Palau that were never available before. One of these days Pat will publish a book on his aerial photography of Micronesia, so keep an eye out..

Bert Yates

Bert came to Palau in 1992 to run the Splash photo center at the Palau Pacific Resort, and later managed NECO Marine dive center. He got his feet wet in US Navy submarines before becoming a PADI dive instructor in the Bahamas. He likes to fiddle with engines and get his hands dirty, but is much more talented with both video and photo, and his amazing work has been published in many magazines and books. Bert left Palau for a year, but couldn't resist the islands, and is now back again working with NECO Marine.

14. Novem: 17 10.
Via Don Padillæ ad Insulas Palaos.
Samal I.
C. S. Guvam
INSULÆ PHILIPPINÆ
Mindanao I.
Carragan
1. Ianu. 17 11.
Via Reditus
25 Decemb.
26. Novemb. 17 10.
Panlog
Panlog inter Palaos Insulas Maxima
A
Sarol
Falivelap
Comul
Ulivela
10. Decemb. 17 10.
I. Sonsorol als S. Andreæ
seu Losacay
Farale seu Poulo I.
Merrieres I. seu Ontocoso
Piguela
INS
Seu
ARCHIP
E Literis
Prov. B
Miss
Linea Æquinoctialis.
Los Papuas
Terra dos Papous
Habitans des Iles Pelew

Insulæ Guaham I. Marianæ
Olutil
Falippe
Tavas
Olor
Faf
Feley
Boo
Ett
Pie
Farvile
Fayo
Old
Pisarap
Lamuet
Pis
Mago
Faravolvol
Falato
Apigap
Yap
Paiz
Olotup
Pug
Loto
B
Carvolvol
Amorsot NB
C
Fataray
Pitamgarus
Falalap
Mayaya
Magomer
Yassor
Luxep
Yaor
Culaot
Eu
Tolobas
Coop
LAOS
87.
Pariali
Sabo
Alangari
Ulimaray
Pogol
NÆ
Corobol
Fagatale
Sabolocolop
ÆPA=
Pirigarava
NOVÆ PHILIPPI=
CAROLINÆ
Aliàs
S. LAZARI vel
AGUS
Otoc
Faratay
Salva
May
Torres
Faramalo
Farali
Faramit
Falu seu
Farilies
Lamureo
Sedes Regis
Tata
Pauli Klain è Soc: JESU è
in Philippinis Insulis
et Theol: Professoris
ilæ; ac ex itinerario
illustris viri Don
Francisci Padilla
Alingaray
Tacaylap
Marioc
Raor
Ulier
Parivu